How Deep Is the Water?

About the principal authors of Real Math™

Stephen S. Willoughby Dr. Willoughby is a professor of mathematics education and mathematics at New York University. He holds an A.B. in mathematics and an A.M.T. in science education from Harvard University and an Ed.D. in mathematics education (1961) from Teachers College, Columbia University. In addition to his many research interests, Dr. Willoughby has taught at all levels from the elementary grades through graduate school and has served on numerous national committees. He has served as president of the National Council of Teachers of Mathematics and is an active member of many other professional organizations. His writings include the books *Contemporary Teaching of Secondary School Mathematics* (1967) and *Statistics and Probability* (1973) and more than fifty other publications.

Carl Bereiter Dr. Bereiter is a professor in the Department of Applied Psychology, Ontario Institute for Studies in Education, and in the Department of Educational Theory, University of Toronto. He received B.A. and M.A. degrees in comparative literature from the University of Wisconsin and, in 1959, obtained a Ph.D. in educational psychology from that university. He was awarded a Guggenheim Fellowship for work on objectives in children's education and was a fellow at the Center for Advanced Study in Behavioral Sciences. Actively involved in research, teaching, and program development in childhood education for twenty years, Dr. Bereiter has published three books, more than fifty articles, and a number of research reports and teaching materials. His books include *Teaching Disadvantaged Children in the Preschool* (1966, coauthored with Siegfried Englemann) and *Arithmetic and Mathematics* (1968).

Peter Hilton Dr. Hilton is Distinguished Professor of Mathematics at the State University of New York (Binghamton) and fellow of the Battelle Seattle Research Center. He holds M.A. and D.Phil. degrees from Oxford University and a Ph.D. from Cambridge University. He has an honorary doctorate of humanities from Northern Michigan University. In addition to his activity in research and teaching as a mathematician, he has had a continuing interest in mathematics education and has served on many national and international committees, including service as chairman of the United States Commission on Mathematics Instruction. Dr. Hilton is a member of the leading mathematics societies. He is the author of several important books and many research articles on algebraic topology, homological algebra, and category theory.

Joseph H. Rubinstein Dr. Rubinstein is a professor of education and chairman of the department of education at Coker College, Hartsville, South Carolina. He received B.A., M.S., and Ph.D. degrees in biology from New York University, completing his studies in 1969. He did basic research in biochemistry and physiology at the Boyce Thompson Institute for Plant Research. He helped develop the Conceptually Oriented Program in Elementary Science (COPES) at New York University. On a postdoctoral fellowship, he spent a year training teachers in New York City. He was director of the Open Court Mathematics and Science Curriculum Development Center for its first seven years. Dr. Rubinstein is the author of papers in elementary science and mathematics education and in research biology, and a contributing author for several elementary curriculum projects.

How Deep Is the Water?

A **Real Math**™ Thinking Story® Book   Level 1

Stephen S. Willoughby
Carl Bereiter
Peter Hilton
Joseph H. Rubinstein

Real Math™ is a product of the
Open Court Mathematics and Science
Curriculum Development Center.

Catherine Anderson, Director

Open Court   La Salle, Illinois

Illustrations
Phill Renaud

OPEN COURT, Thinking Story, and ✲ are registered in the U.S. Patent and Trademark Office.

Copyright ©1985, 1981 Open Court Publishing Company

All rights reserved for all countries. No part of this book may be reproduced by any means without the written permission of the publisher.

Printed in the United States of America

ISBN 0-89688-602-6

Contents

	Introduction	vii
1	How Many Piglets?	2
2	Willy in the Water	6
3	Mr. Sleeby's Party	10
4	"It's Not So Easy," Said Mr. Breezy	14
5	Mr. Sleeby Goes Shopping	18
6	Exactly What to Do	22
7	Willy Looks in the Mirror	26
8	Mr. Mudanza Builds a Better Tree	30
9	Mark Builds a Birdhouse	34
10	Mr. Sleeby Buys a Candy Bar	38
11	Mrs. Nosho's Fish Stories	42
12	Manolita's Magic Number Machine	46
13	Mr. Mudanza Makes Lunch	50
14	Manolita's Magic Minus Machine	54
15	How Mrs. Nosho Doubled Her Money	58
16	The Third House on Fungo Street	64
17	The Lemonade War	70
18	Mr. Mudanza Changes Houses	76
19	Trouble in the Garden	80
20	How Deep Is the Water?	86
	Correlation of the Thinking Story® Book and Teacher's Guide for Real Math™, Level 1	93

Introduction

Nature and purpose of the Thinking Story® book

The Thinking Story® book is an essential part of the Open Court Real Math™ Program. It is aimed at developing *quantitative intelligence*—creativity and common sense in the use of mathematics. The thinking skills that are stressed include choosing appropriate operations, recognizing when a mathematical model is or is not appropriate, recognizing absurd answers, recognizing obvious answers (those that don't need calculation), and solving realistic measurement problems.

The various characters in the stories and problems have peculiarities that the children come to know: Mr. Sleeby, for example, is always forgetting things; Ferdie is always jumping to conclusions without thinking; Mr. Breezy is always giving more information than is needed and so makes easy problems seem difficult. When the children approach a certain story or problem, they do so not with an attitude of "This is a story in which I have to add," but rather one of "This is a story in which I have to pay attention to what numbers are important." In this way they become used to thinking rather than only to carrying out arithmetical operations mechanically. Because the stories and problems are filled with surprises, the children can never fall into a comfortable, nonthinking rut.

The stories are real stories, designed to be read to the children by an adult. As a story unfolds, the children are asked questions that prompt them to think *ahead* of the characters—to spot what is wrong with what a character has done or said, to anticipate what is going to happen as a result, or to think of other possibilities that the character hasn't considered. The thinking problems that follow the story are in the same vein, but much shorter.

Use of the Thinking Story® book*

Scheduling In its lesson plans the first-grade Teacher's Guide suggests which of the Storybook selections might be appropriate. In general, provision is made for one story, whether new or repeated, each week. New and repeat problems—usually two or three of each—are also scheduled in the Teacher's Guide, on days when no stories are read. (See the Correlation of the Thinking Story® Book and Teacher's Guide on page 93 of this book.) You may want to change the numbers a little when repeating a problem. Checking off each problem as you use it might also be helpful.

Method of presentation Read the stories aloud and do the problems with the group as a whole, unless you have an aide who can work with smaller groups.

Here are a few suggestions:

Use pacing and emphasis to facilitate thinking. Read each story clearly and methodically enough to let the points sink in. Give the children time to think about each question, but not so much time that they forget the point of the story. Make it clear, by voice and by eye contact, whether you are reading narrative or directing a question to the children. The text helps in this by setting off in boldface type the questions to be asked. Emphasize, by a pause and a fresh start, when you are leaving one point and shifting to another one in a story. Otherwise the children may think you are responding to their answers when you are actually going on to a new event.

Stay on the track. Many of the story questions *could* be discussed at length but shouldn't be, because the children would lose the thread of the story. The questions are written for brief answers that move the children on to the next point in the story rather than to related issues. Even if children at first show an inclination to go off on tangential discussions, they will soon come to prefer getting on with the story, as long as the reader doesn't encourage such digressions. Freewheeling discussion is fine, but these stories are not a good vehicle for it.

Discuss problems. Always ask the children how they figured out their answers. Allow debate. Some of the given answers are arguable, and the arguments can be very productive. For instance, one problem states that Ferdie has 15 cents. His mother gives him a dime to buy a newspaper for her. How much money does Ferdie have left after he buys the newspaper? The intended answer, of course, is 15 cents. But a child might argue that the paper cost more than 10 cents. All right, what would happen then? What would Ferdie do? What would his mother do? Working through such arguments can provide good mathematical thinking.

Think along with the children. Try to maintain an attitude which shows that the important thing is to think carefully about the problems, not to know the answer instantly.

Use response cubes or finger sets. Many of the problems call for numerical answers, and response cubes or finger sets are the best way to get all the children to respond, rather than just the one you happen to call on.

Use expanded numbers when appropriate. The numbers in the Storybook are all given in conventional form. During the period when the children are using expanded notation (lessons 43–119), the numbers in the stories should be read in expanded form. That is, the numbers between ten and a hundred should be read as "ten," "ten and one," "ten and two," "two tens," and so on, through "ten tens." This form makes it easier for the children to grasp the decimal structure of our number system.

Make the children feel good about their answers. Don't praise children only when they give the answers you happen to have in mind. Recognize that for most questions there are a

---

*If you are not using the Open Court Real Math Program, some of these suggestions will not be applicable; however, most of those concerning methods of presentation will be useful in any classroom or home situation.

Introduction

number of possible answers, that answers may be sensible even if they are not quite correct, and that even a clearly incorrect answer probably entailed some worthwhile thinking on the child's part. Try to show a lively interest in both the questions and the answers, whatever they may be.

Don't encourage snobbery. The characters in these stories are well aware of one another's foibles. They are all warm-hearted people who support each other and treat each other with respect. The stories are written so as to convey this attitude to the children as well. A lapse in thinking is not necessarily more contemptible than a broken arm, although both are handicaps that one does well to recognize and avoid if possible. So don't ridicule the characters or make a point of the children's intellectual superiority over them.

Keep a light touch. The stories are often farcical. Many of the questions call attention to something amusing. The stories are not intended to have any drama or serious message, so don't try to invest them with either.

Adjust to individual and group differences. Don't expect every child to get the problems and the story questions the first time around. That is why there is repetition. Many more children will get the point and will reason correctly the next time. The problems have a mixture of difficulty levels. It shouldn't be necessary to select, but you might call on slower children for answers to the more obvious questions.

Problems and stories get harder toward the end of the book. A slower first-grade group might not be able to make it all the way, but instead could profit from more reruns of the earlier material. With a faster group, go through the whole book (even the early parts are sufficiently challenging the first time through), but don't do much repetition of the earlier stories and the easier problems.

Principal characters in the Storybook

The main characters in *How Deep Is the Water?* are special. Each has a peculiarity in his or her thinking. The children learn to recognize these peculiarities and to avoid them in their own thinking.

Mr. Sleeby is always forgetting things. The children learn to keep in mind the kinds of things that Mr. Sleeby forgets.

Ferdie, overconfident and impulsive, is always leaping to conclusions. The children learn to consider the facts that Ferdie ignores.

Portia, Ferdie's younger sister, is more cautious. She does not jump to conclusions, and thus often provides a balance to Ferdie's impulsiveness.

Mrs. Nosho is so vague that people have trouble understanding what she is talking about. The children learn how to say things more clearly and to ask the kinds of questions that are needed to find out what Mrs. Nosho means.

Mr. Breezy usually says too much. In trying to be helpful, he confuses people with irrelevant details. The children learn to distinguish the essential from the irrelevant in what Mr. Breezy says.

Mark, Mr. Breezy's son, is always asking questions that often help people clear up their problems.

Manolita thinks that everything happens by magic. The children learn to figure out how things really happen.

Mr. Mudanza, Manolita's father, always changes things "a little." (Given a new dresser, he throws away everything but one drawer and turns that into a wagon.) The children learn to perceive, through mental imagery, the results of Mr. Mudanza's changes.

Willy only wishes for what he wants to happen. The children learn to think of ways to make things really happen.

Introduction

Metric units in the Storybook

Metric units are the only standard units of measure used in *How Deep Is the Water?* They are included not to teach the metric system but to help give the children a feel for how some of the common metric units are used in everyday activities.

In a few of the stories and problems you'll be asking the children to show with their fingers or hands the size of an item such as the Noshos' 4-centimeter-thick rug. Thus you should know the approximate magnitudes of the metric units used in the Storybook. The following table is a handy reference.

	Unit	Symbol*	Relation to Common Objects or Events
Length	centimeter	cm	A paper clip is about 1 centimeter wide and 3 centimeters long.
			An unsharpened pencil is about 20 centimeters long.
			This line is 1 centimeter long: ─── .
	meter	m	A full-size automobile is about 2 meters wide.
			Most classroom doors are about 1 meter wide.
			A telephone pole is about 10 meters high.
	kilometer	km	It takes about 12 minutes for an adult to walk 1 kilometer.
			There are about 4500 kilometers between Los Angeles and New York (driving distance).
Weight	gram	g	Two paper clips weigh about 1 gram.
			A nickel (or a marble) weighs about 5 grams.
	kilogram	kg	The average weight of a newborn baby is 3–4 kilograms.
			A pair of size-10 men's shoes weighs about 1 kilogram.
Volume	liter	L	Four average-size glasses hold about 1 liter of liquid.
			The gas tank of a full-size automobile holds about 90 liters.

*The same symbol is used for the singular and plural forms.

Introduction

How Deep Is the Water?

How Many Piglets?

1
How Many Piglets?

Note: The illustration that goes with this story is best shown at the end of the reading, as a picture puzzle: Find all the piglets. However, if children don't know what baby pigs look like, they had better see the illustration at the start.

Ferdie and Portia could hardly wait for Saturday to come. They were going out to Grandfather's farm to see Martha's babies. Martha is Grandfather's pig, and these were her first babies. Ferdie and Portia had never seen piglets before.

What do you think piglets are?

When Saturday came, Ferdie and Portia and their mother got on the bus and rode out to Grandfather's farm. The bus driver let them out right at the gate. Ferdie and Portia ran ahead to the barnyard, where they found Grandfather standing by the pigpen. Martha was standing in the pen, eating—as usual. She didn't even look up. She is not the world's friendliest pig.

What would you have expected the world's friendliest pig to have done?
Would she have smiled at Ferdie and Portia?

All around Martha, running this way and that, were her piglets. Some were pink and some were black, and some were partly pink and partly black.

"Look at them run!" said Ferdie.
"How many piglets are there?" asked Portia.
"Count them yourselves," said Grandfather with a smile, "if you can."
"Of course I can count them," said Ferdie. "That's easy."
Ferdie crouched down beside the pen and counted the piglets as they ran past. He counted, "1, 2, 3, 5 . . ."
"You made a mistake," said Portia.

What mistake did Ferdie make? (He skipped 4.)
What should he have said? ("1, 2, 3, 4, 5 . . .")

"You skipped 4," said Portia.

"All right," said Ferdie, "I'll start again."

This time he didn't skip any numbers. Every time a piglet ran past, he counted. He counted, "1, 2, 3, 4, 5, 6, 7, 8, 9, 10." Then he shouted, "Ten piglets! That's a lot!"

"H'm," said Grandfather, "I didn't think there were that many."

Could Ferdie have made a mistake? How?

"I think you counted some piglets more than once," said Portia. "You counted every time a piglet ran past, and some of them came past more than once. Let me try."

Portia looked into the pen, where the piglets were still running around. She said, "There's a pink one. That's 1. There's a black one. That's 2. There's a spotted one. That's 3. And, oh, there's one with a funny tail. That's 4. Martha has 4 piglets."

"You did that wrong," said Ferdie. "You didn't count all the piglets."

How could Portia have made a mistake?

"You counted only 1 pink one," said Ferdie, "and there's more than 1 pink one. See? And there's more than 1 black one, too. I don't know how many piglets there are. I wish they'd stand still so we could count them."

"Just wait," said Grandfather. "Maybe they will."

In a little while Martha finished eating and lay down on her side. The piglets stopped running around. They went over to their mother and started feeding.

"Now we can count them," said Portia. "They're all in a row." She counted, "1, 2, 3, 4, 5."

How many piglets did she count? (5)

"Martha has 5 piglets!" said Portia.

"That's strange," said Grandfather. "I thought she had more. But you're right, there are only 5 piglets there."

Just then they heard a sound, "Eee, Eee, Eee," and another piglet that had been off by itself came running across the pen and joined the others.

How many piglets are there now? (6)
How do you know?

How Many Piglets?

Problems

1. Portia got a new coat. Before, she had only a brown coat. Now she has a green coat too.

How many coats does Portia have? (2)

2. Count how many things Mark did: First he washed his hands. Then he washed his face. Then he brushed his teeth. Then he combed his hair.

How many things did Mark do? (4)

3. Mrs. Nosho had 7 rosebushes growing in her backyard. She picked 1 rose off each bush and gave them all to Willy.

How many roses did Willy get? (7)

4. Manolita counted all the fingers on one hand, but she didn't count the thumb.

How many fingers did Manolita count? (4)

5. "Oh, every wheel on my tricycle is broken," moaned Ferdie.

How many wheels are broken? (3)

6. Mr. Breezy likes radios. He has a radio in the kitchen. He has a big radio in the living room. He has a clock radio in his bedroom. And he has a radio in his car.

How many radios is that? (4)

7. Figure out how many horses Grandfather has: Grandfather has a horse that he rides when he wants to go horseback riding. He uses the same horse to pull a wagon when he does farm work. That same horse, whose name is Arnold, sometimes pulls a sleigh in the winter. Grandfather takes good care of Arnold, because Arnold is the only horse he has.

**How many horses does Grandfather have? (1)
How do you know for sure?**

8. Loretta the Letter Carrier was delivering the mail. Count how many letters she delivered: First she delivered a letter to Mr. Mudanza. Then she took a letter to Mark. Then she took a letter to Mrs. Nosho. Then she delivered 1 letter to Ferdie and 1 to Portia.

How many letters did Loretta deliver altogether? (5)

9. Mark has 6 big cans of paint that he uses to paint model airplanes. He painted one airplane white. He painted another airplane red. And he painted another airplane yellow.

How many cans of paint does Mark have? (6)

How Many Piglets?

Willy in the Water

2

Willy in the Water

Willy the Wisher is always wishing that things were different, but he doesn't know what to do to make them different. One time he was on vacation at the seashore. Willy had great fun collecting things along the beach and wading in the water, but he kept wishing things were a little different.

Willy had 3 sea shells that he had picked up along the shore. "I love these pearly shells," he said, "but I wish I had 4 of them instead of 3."

What could Willy do to have 4 shells instead of 3? (find another)

Just then Willy happened to notice another shell on the sand. He picked it up. Then he counted his shells again.

How many shells does he have now? (4)
How do you know?

"My wish came true," said Willy. "I have 4 shells now." He also had 4 shiny white stones that he had found. But that was not how many stones he wanted.

"I wish I had 2 shiny stones instead of 4," said Willy.

What could Willy do to have 2 stones instead of 4? (throw or give 2 away)

Willy couldn't think of any way to have 2 instead of 4, so he kept carrying those 4 heavy stones with him until they made a hole in his pocket and 2 of them fell out.

How many stones did Willy have then? (2)

"My wish came true again," said Willy. "I have 2 shiny white stones now, just the way I wanted." But his wish didn't stay true for long, because the other 2 stones fell out through the hole in his pocket too, and then he didn't have any.

Willy had also found 5 crab claws, but he wished he had 7.

What could Willy do to have 7 crab claws instead of 5? (find 2 more)

Willy was standing in the water near shore, wishing he had 7 crab claws. Then he noticed that the water was about 8 centimeters deep where he was standing.

How deep is that? Show with your hands. [Demonstrate the correct depth.]

"The water feels good on my ankles," said Willy. "But I wish it was 10 centimeters deep instead of 8."

What could Willy do about it? (wade in deeper)

Willy didn't think about wading in deeper. He just stood there in the water, wishing. The tide was going out, so the water kept getting lower and lower. Soon the water wasn't 8 centimeters deep any more, but 4 centimeters, then 2 centimeters, then 1 centimeter deep, and finally there wasn't any water around Willy's feet at all.

"Sometimes my wishes come true and sometimes they don't," said Willy.

Willy in the Water

Problems

1. Willy's uncle gave him a dog. After a while the dog had 3 puppies.

   How many dogs did Willy have then? (4)

2. "I'm 6 years old," said Willy. "When I have my next birthday, I'll be 5."

   What's wrong with what Willy said?

3. Yesterday Ferdie counted 6 ducks in the pond. Today there are 8.

   What could have happened? (More ducks arrived; 2 eggs hatched.)

4. Portia had 3 nickels. She spent 2 for some candy.

   How many nickels does she have left? (1)

5. Mark's mother gave him 5 grapes. He lost 1 and ate the rest.

   How many grapes did he lose? (1)

6. Mark went shopping with his mother, and they bought 2 pairs of shoes. One shoe slipped out of the box before they got home, but they didn't see it fall.

   How many new shoes did they have when they got home? (3)

7. Portia had 1 cookie. Mr. Sleeby gave her 2 more, but she gave 1 of them to Ferdie.

   How many cookies did she have then? (2)

8. Manolita bought 5 notebooks at the store. Then she found out that she needed only 2, so she took the others back.

   How many notebooks did she take back? (3)

Mr. Sleeby's Party

3

Mr. Sleeby's Party

Portia and Ferdie stopped by to visit Mr. Sleeby. "I'm getting ready to have a party," he said. "You can help me by setting out these funny hats for people to wear."

Portia and Ferdie counted the funny hats. There were 4 little hats and 3 big hats. Ferdie said, "I'll bet I can figure out what kind of people will be at your party: 4 people with little heads and 3 people with big heads!"

"That's good thinking," said Mr. Sleeby. "You almost have it figured out."

"I have an idea," said Portia. "Are you inviting some children and some grownups?"

"Yes, I am," said Mr. Sleeby.

**Can you figure out how many children and how many grownups are invited to the party? (4 children, 2 or 3 grownups)
Are you sure?**

"I know! I know!" shouted Ferdie, who always liked to be first with an answer. "Four children and 3 grownups. Who are they?"

"Let me try to remember," said Mr. Sleeby. "The grownups are Loretta the Letter Carrier and Mr. Mudanza and Mrs. Mudanza."

Portia counted the 3 big hats again. "Mr. Sleeby," she said, "there's going to be 1 grownup at your party who isn't going to have a funny hat to wear."

**Who is that? (Mr. Sleeby)
What did Mr. Sleeby forget? (to count himself)**

Mr. Sleeby said, "Yes, I forgot about myself. I had an invitation all ready to send myself, but I forgot to mail it."

"Who are the children you're inviting?" Ferdie asked.

"There's Willy and Manolita and some other children," said Mr. Sleeby. "I can't remember their names right now."

**How many other children should there be? (2)
How do you know?**

"There should be 2 other children," Ferdie said. "I can tell because there are 4 hats, and Manolita and Willy and 2 more make 4."

"Are the other 2 children by any chance a brother and a sister?" Portia asked.

"Yes, I believe they are," said Mr. Sleeby.

"And are their names Ferdie and Portia?" asked Ferdie.

"That's right," said Mr. Sleeby. "What clever children you are!"

"Oh, good!" said Portia. "We're invited to the party! When is your party, by the way?"

"It's right now," said Mr. Sleeby. "I believe I see some guests coming up the sidewalk. You children can help me by standing at the door and handing out funny hats to all the people when they come in."

Do you remember how many funny hats there are? (7)
How many big ones? (3)
How many little ones? (4)

First Mrs. Mudanza walked in. They gave her a funny clown's hat. Then Mr. Mudanza came in. They gave him a funny hat with a feather in it. He changed it a little by tying the feather in a knot, and said, "Thank you very much."

How many big hats are left? (1)

"There's 1 big hat left," said Portia. "Oh, I know what to do with it. Here, Mr. Sleeby." She put the last big funny hat on Mr. Sleeby's head, and he went dancing around the room.

Next they heard a child's voice outside, saying, "I wish the door was open so I could come in."

Can you guess who that might be?

"That sounds like Willy the Wisher," said Ferdie. "Come in, Willy." They gave him one of the little hats.

Then they heard another voice outside, saying, "Door, door, open wide!"

"That sounds like Manolita," said Portia. Sure enough, there was their friend Manolita, standing at the door. They let her in and gave her a little hat.

How many little hats are left? (2)

Ferdie and Portia waited, but no more children came to the door. "I guess some children that you invited aren't coming," said Ferdie.

Was he right? (no)
What had he forgotten?

"You forgot to count yourselves," said Mr. Sleeby. He put the last 2 funny hats on Ferdie's and Portia's heads, and they all danced around the room.

Suddenly there was a knock at the door. It was Loretta the Letter Carrier. "I hope I'm not too late for the party," she said. "I just finished delivering the mail."

"You're not too late," said Portia, "but something is wrong. There are no hats left. There's no funny hat for you to wear."

Why not? Can anyone remember why there aren't enough hats?

"It's all right," said Loretta. "I have my letter carrier's cap, and it will do."

"It will do if we change it a little," said Mr. Mudanza. He turned Loretta's hat around and put the feather from his own hat in it.

"Now we all have funny hats," said Portia, and the party began.

Mr. Sleeby's Party

Problems

1. Loretta the Letter Carrier and her husband, Roger, have a baby and a dog. Loretta's mother lives with them.

How many people live in their house? (4)

2. Portia gave a party. She invited 2 girls and 2 boys. They all came.

How many girls were at the party altogether? (3)

3. "I had 6 carrots this morning," said Mr. Sleeby. "Then I ate 1 or 2 of them—I can't remember which."

How many carrots does he have left? (4 or 5—can't tell.)
What do you need to know to be sure? (how many he ate)

4. Mark had 5 apples. Then he ate 2 apples, only they were someone else's apples, not his own.

How many apples did he have left? (5)

5. Manolita walked 2 blocks to the store. Then she walked a block to Mark's house. After they played a while, she walked a block to get to her own house.

How far did she walk altogether? (4 blocks)

6. Portia planted 4 beans, but 3 of them didn't grow.

How many bean plants grew? (1)

7. Mr. Mudanza wanted to send his brother the smallest letter in the world. He made an envelope just the size of a postage stamp.

How big is that? Show with your fingers. [Demonstrate the correct size.]

He wrote his brother's address on the front of the envelope. Then he pasted a stamp on the front of the envelope and mailed it. His brother never got the letter.

Why not? (The stamp covered the address.)

8. Willy's father gave him 5 peanuts. Willy lost 1 and ate the rest.

How many peanuts did he eat? (4)

9. When Mr. Sleeby got back from a trip, there were 7 old newspapers on his porch. He picked up 2 of them.

How many old newspapers did he leave on his porch? (5)

Mr. Sleeby's Party

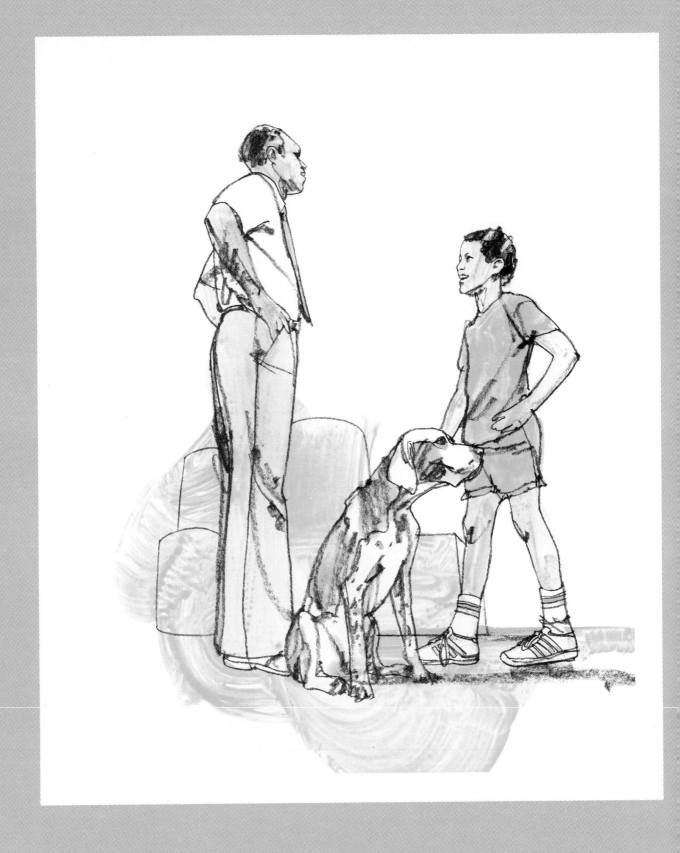

"It's Not So Easy," Said Mr. Breezy

14

4
"It's Not So Easy," Said Mr. Breezy

Mr. Breezy runs a training school for dogs. People send their dogs to him, and he teaches the dogs to obey commands and sometimes to do tricks. Mark likes to help his father at the dog-training school.

"Any jobs for me today?" Mark asked.

"I have some work that needs to be done, but it's not so easy," said Mr. Breezy. "The first job is to figure out how many cans of dog food we have left."

"I'll go to the storeroom and start counting," said Mark.

"It's not that easy," said Mr. Breezy. "You'll find there are 10 cans of dog food there altogether. But 6 of the cans are right-side up. And 3 cans are upside down. Oh, 1 can is on its side. Do you think you can handle all those numbers?"

"I don't need to," said Mark. "I think I know the answer already."

How can Mark know the answer without going to the storeroom?
How many cans of dog food are there altogether? (10)
What about the 6 cans that are right-side up and the 3 that are upside down and the 1 that is on its side?

"There are just 10 cans," said Mark. "I don't have to go to the storeroom, because you told me how many at the beginning. It doesn't matter how many are right-side up and upside down and on their sides."

"You're pretty good with numbers," said Mr. Breezy. "I'm proud of you. But here's a problem that's not so easy. You know the chains we use to lead the dogs with when we're teaching them to follow us? Well, I'm trying to figure out how long the chains should be."

"Maybe we can find a book on dog training that will help," said Mark.

"I've already looked it up in a book," said Mr. Breezy, "but the book gives so many different numbers that there's no way to know which number is best."

Mr. Breezy started to read from the book: "A good chain should weigh about 2 kilograms, should be made of wire 1 centimeter thick, should be 5 meters long, should have about 25 links to the meter, and should be shiny."

"I told you this wouldn't be easy," said Mr. Breezy.

"I think I have the answer already," said Mark.

How long should each chain be? (5 m)
What about all the other numbers?

"The book says the chains should be 5 meters long," said Mark. "The rest of those numbers don't tell how long they should be. They tell you other things about them."

"Spoken like a true Breezy," said Mr. Breezy. "Since you're so good with numbers, maybe you can help with a really tough problem. We need to buy more pens, because we can keep only 1 dog in each pen. We have so many dogs here now that there aren't enough pens for all of them. What is not so easy is to figure out exactly how many new pens we need."

"How many pens do we have now?" Mark asked.

"I don't know," said Mr. Breezy, "but they're all full."

"Then how many dogs do we have here?" Mark asked.

"Let me think," said Mr. Breezy. "We have 6 dogs in the pens out back, 4 dogs in the pens downstairs, 5 dogs in the pens upstairs, and 3 dogs in no pens at all."

"And you want to know how many more pens we need?" Mark asked.

"I knew you wouldn't think this one is easy," said Mr. Breezy. "So how will we get the answer?"

"But I already have the answer!" said Mark.

How many more pens do they need? (3)
How can you tell?

"We need 3 more pens," Mark told his dad, "for the 3 dogs that aren't in any pens now."

"You make it all sound easy!" exclaimed Mr. Breezy. "What's your secret?"

"My secret," said Mark, "is to pay attention only to the numbers that matter."

How can you tell what numbers matter?

"It's Not So Easy," Said Mr. Breezy

Problems

1. Mr. Breezy said, "Our cat, Abigail, had 2 kittens last year. This year she had 3 kittens. How many feet does Abigail have altogether?"

How many? (4, like any other cat)

2. Mark got up early to mail a letter. Three people saw him while he walked to the mailbox. Nobody saw him while he walked home.

How many people saw him walking altogether? (3)

3. Ferdie was doing dishes. He had 10 plates left to wash. He washed 3 spoons.

How many plates did he have left to wash? (10)

4. Manolita had 3 sheets of paper, and she made each sheet into a paper airplane. Two of the planes didn't fly right, so she threw them away. Another one got wet, so she threw it away too.

How many planes did she have left? (zero)

5. Ferdie doesn't always pay attention to what people tell him. His mother asked him to buy 6 doughnuts at the store, but he bought only 5. His mother sent him back for more.

How many more doughnuts did Ferdie have to get? (1)

6. Mr. Breezy said, "I earned 3 dollars, then I caught 2 fish, then I caught a cold, then I lost all my money, then I ate 2 walnuts. How many dollars did I have left?"

How many? (zero)

7. Willy lives with his father, his grandmother, his sister, and his brother. Willy and his grandmother were sick on Tuesday, but everyone else felt all right.

How many felt all right on Tuesday? (3)

8. The teacher said, "Now open your books to page 8." Mark opened his to page 10.

Should Mark turn toward the beginning or toward the end of the book? (the beginning)

"It's Not So Easy," Said Mr. Breezy

Mr. Sleeby Goes Shopping

5

Mr. Sleeby Goes Shopping

Mr. Sleeby invited some friends to his house for lunch. Then he looked in his refrigerator and saw that he didn't have enough food, so he figured out just what he needed to buy and made a list. The list said: "6 eggs, 7 apples, 5 oranges, 8 tomatoes." Then Mr. Sleeby left the list on the kitchen table and went to the store.

"Good morning, Mr. Sleeby," said Mrs. Ling, who owns the grocery store. "What do you need today?"

"Well," said Mr. Sleeby, "I need eggs, apples, oranges, and tomatoes."

"How many eggs?" asked Mrs. Ling.

"I don't know for sure," said Mr. Sleeby.

Why isn't Mr. Sleeby sure how many eggs he needs? (He forgot his list.)

"Oh, I see you forgot your list again," said Mrs. Ling. "Can you remember *about* how many eggs you need?"

"All I can remember," said Mr. Sleeby, "is that it is 1 more than 5."

Can you figure out how many eggs Mr. Sleeby needs? (6)

"I believe you need 6 eggs," said Mrs. Ling. She gave Mr. Sleeby a half-carton of eggs with 6 eggs in it.

"Now," said Mrs. Ling, "how many apples do you need?"

"I can't remember that either," said Mr. Sleeby, "but I know it is less than 10."

Can Mrs. Ling be sure exactly how many apples Mr. Sleeby needs? Why not?

"You haven't told me enough, Mr. Sleeby," said Mrs. Ling. "There are lots of numbers that are less than 10. Is it 8 apples you need?"

"No," said Mr. Sleeby. "It is 1 less than 8."

Now can Mrs. Ling be sure? (yes)
How many apples does Mr. Sleeby need? (7)

"You need 7 apples," said Mrs. Ling. "Here they are. Now, how many oranges do you need?"

"I don't know," said Mr. Sleeby, "but it's the same as the number of fingers I have on one hand."

**How many oranges does Mr. Sleeby need? (5)
How do you know?**

"That was easy," said Mrs. Ling. "Here are your 5 oranges, 1 for each finger. Now all we need to know is how many tomatoes you need."

"I really can't remember that," said Mr. Sleeby. "All I know is that I'm going to use 8 tomatoes for lunch, and I don't have that many tomatoes in my refrigerator."

**Can Mrs. Ling be sure how many tomatoes Mr. Sleeby needs to buy? (no)
What else does she need to know? (how many are in his refrigerator)**

"I can't be sure how many you need," said Mrs. Ling. "If you have a lot of tomatoes in your refrigerator already, then you won't need to buy very many. I know you want to have 8 tomatoes altogether. It would help if I knew how many tomatoes you have in your refrigerator right now."

"Oh, I can tell you that," said Mr. Sleeby. "I don't have *any* tomatoes at home."

Now can you figure out how many tomatoes Mr. Sleeby needs to buy? (8)

"I think I can figure that one out," said Mrs. Ling. "You need 8 tomatoes and you don't have any, so you need to buy all 8 tomatoes. Well, here they are. I hope we didn't forget anything."

"I don't think we did," said Mr. Sleeby. "I can't remember anything we forgot."

Mr. Sleeby Goes Shopping

Problems

1. Somebody asked Mr. Sleeby, "How many children do you have?" "I don't remember," said Mr. Sleeby, "but it's 1 less than the number of noses I have."

How many children does Mr. Sleeby have? (zero)
How did you figure it out?

2. "How old are you?" somebody asked Portia. Portia pointed to her ears, her eyes, her nose, and her mouth. "Count them all up," she said. "That's how old I am."

How old is she? (6)

3. Yesterday Ferdie borrowed 3 pennies from Manolita. Today he gave back 2 pennies.

Does he still owe Manolita any money? (yes)
How much ? (1¢)

4. The Noshos' dog, Muffin, buried 2 bones, then 3 cans, then 1 potato, then 2 bones, then 3 old newspapers.

How many bones did Muffin bury altogether? (4)

5. "We've had bad weather for 5 days," said Loretta the Letter Carrier. "It rained for 2 days and snowed the rest of the time."

How many days did it snow? (3)

6. Willy has a rabbit that is worth a dollar. His brother has 2 rabbits just like it.

How much are his brother's rabbits worth altogether? ($2)

7. Portia ate 2 apples and an orange after lunch. Later she ate another apple.

How many oranges did she eat altogether? (1)

8. The puddle in Willy's backyard is 4 meters wide. It used to be 5 meters wide.

What could have happened? (It might have dried up, drained, and so on.)

9. Ferdie and Portia each have a pillow stuffed with feathers. Portia thinks her pillow has more feathers in it, and Ferdie thinks his does. "I know how we can find out," said Ferdie. "Let's open up our pillows and count all the feathers."

Is that a good idea? Why not? (Too many feathers; they'd blow all over.)
Can you think of a better way for them to tell without counting? (by weighing the pillows or measuring their thickness, for instance)

Mr. Sleeby Goes Shopping

Exactly What to Do

6

Exactly What to Do

"It's time to walk the dogs," said Mr. Breezy.

That was a job Mark sometimes did at his dad's training school for dogs. He took the dogs to a small park that was right next door to the school.

"How many times should I walk around the park with the dogs tonight?" Mark asked.

"That's not so easy to figure out," said Mr. Breezy. "But here's exactly what you'll need to do. First pat each dog on the head 2 times. Then walk the dogs around the park 2 times. Then stop at the drinking fountain for 3 drinks of water. Then walk the dogs around the park 1 time. Then say 'Nice doggie' 2 times. Then put the dogs back in their pens."

"I see what you mean," said Mark. "That's not so easy, but at least now I know how many times to walk around the park."

How many times does Mark need to walk the dogs around the park? (3) [Repeat Mr. Breezy's instructions, if necessary.]

Mark figured out that his dad wanted him to walk the dogs around the park 3 times. He knew that patting the dogs on their heads and saying "Nice doggie, nice doggie" wouldn't help him figure out how far to walk.

When Mark brought the dogs back from their walk, he remembered that one dog, Bowser, hadn't been fed yet. "How much dog food should I give Bowser?" Mark asked.

"That question's not so easy," said Mr. Breezy. "Bowser's pretty fussy about how much he eats, so you have to do everything exactly right."

Does that help Mark to know how much food to give Bowser? (no)

"Okay," said Mark. "I'll listen very carefully."

"Here's exactly what to do," said Mr. Breezy. "Take 2 spoons. Then put 1 spoon in the sink. Then take the other spoon and put 2 spoonfuls of dog food in the bowl. Then say 'Good old Bowser!' 3 times. Then put 2 spoonfuls of water on the flowers on the window sill. Then mop the floor. Then put another spoonful of dog food in the bowl, and then another one. Then watch the news on television. Then put 1 more spoonful of water on the flowers."

"Now I know exactly how much food to give Bowser," said Mark.

Do you know how much food Bowser should get? (4 spoonfuls)

Mark did everything exactly the way his dad had told him to. It took almost an hour until he finished feeding Bowser.

"You're quick at figuring things out," said Mr. Breezy, "but you don't seem to move very fast. How could it take you almost an hour to feed a dog 4 spoonfuls of dog food?"

Why did it take so long?
What other things did Mark have to do?
(take 2 spoons, put 1 in the sink, say "Good old Bowser!" 3 times, put 2 spoonfuls of water on the flowers on the window sill, mop the floor, watch the news on television, and put 1 more spoonful of water on the flowers)

Exactly What to Do

Problems

1. "It's not so easy," Mr. Breezy told Mark, "but here's what you'll need to do: The first thing is to put on your shoes. The third thing is to bring in the newspaper. The fourth thing is to pet the dog."

What did Mr. Breezy forget? (the second thing)

2. Yesterday it rained for 2 hours before lunch. Then the sun came out for a while. Then it rained for another hour.

How long did it rain altogether? (3 hours)

3. Mr. and Mrs. Nosho have 3 children. Two have gone on a bus to visit their grandmother. One is away at college.

How many children are at home? (zero)

4. Mr. Sleeby has 5 fishbowls. Each bowl has 0 fish in it.

How many fish does Mr. Sleeby have altogether? (zero)

5. Mr. Breezy made up this riddle: "I'm thinking of a number that's less than 7 and more than 2 and less than 4 and more than 1."

Can you figure out what number it is? (3)
How can you be sure? (It's the only number that's more than 2 and less than 4.)

6. Mr. Breezy and Mr. Nosho and Mr. Mudanza were having coffee at Mr. Sleeby's. "I'm going to have a piece of pie," said Mr. Sleeby. "How about the rest of you?"

"Yes, thank you," said Mr. Breezy.
"Yes, thanks," said Mr. Nosho.
"No, thank you," said Mr. Mudanza. "I'm on a diet."

How many pieces of pie should Mr. Sleeby serve? (3)

7. Mr. Mudanza had a wooden chair with 4 legs. Then he changed it a little. He took a broomstick and cut it in half. He put each half under the chair for an extra leg.

Get a picture of that chair in your mind. How many legs does it have? (6)

8. Mark's mother likes rings. Mark wants to buy her a ring for her birthday. He wants to be sure it's the right size for her finger, but he can't ask her—he wants it to be a surprise.

Think of ways that Mark could find out what size ring to buy. (measuring one of her rings; making a clay model of it; finding a stick that fits into it; trying it on his thumb; and so on)

9. Mr. Breezy said, "Our cabin was full of flies. I swatted 6 flies with a newspaper. Then I swatted 4 more with a magazine. Then I got 2 of them with a dustcloth and 3 of them with a bath towel. But there were still a lot of flies left."

What worked best for swatting flies—paper or cloth? (paper)
Do you want to hear the problem again?

Exactly What to Do

Willy Looks in the Mirror

7

Willy Looks in the Mirror

Remember how Willy the Wisher is always wishing things were different? One day Willy was looking in a mirror. He saw someone in the mirror who looked just like him. Willy didn't want the reflection to look just like him. He wished it looked different. He wished it had 1 more eye than he did.

How many eyes did Willy wish the reflection had? (3)

He wished it had twice as many ears as he did.

How many ears did Willy wish the reflection had? (4)

The reflection showed 4 books, the same number Willy was carrying. "I wish it showed 2 more books," said Willy.

How many books did Willy wish the reflection had? (6)

Willy's nose was 3 centimeters long. He wished the one in the reflection was a centimeter shorter.

How long did Willy wish the nose in the reflection was? (2 cm)

Willy also wished that the reflection had 1 more finger on one hand and 1 less finger on the other hand.

How many fingers did Willy wish the reflection had on each hand? (6 on one, 4 on the other)

"I wish that reflection would at least smile a little bit and look friendly," said Willy.

How could Willy get the reflection to smile?

Willy thought he saw a little bit of a smile on the face in the mirror, so he smiled back. Then the reflection gave him a big, friendly smile. "He's not such a bad guy after all," said Willy. "I guess I like him the way he is."

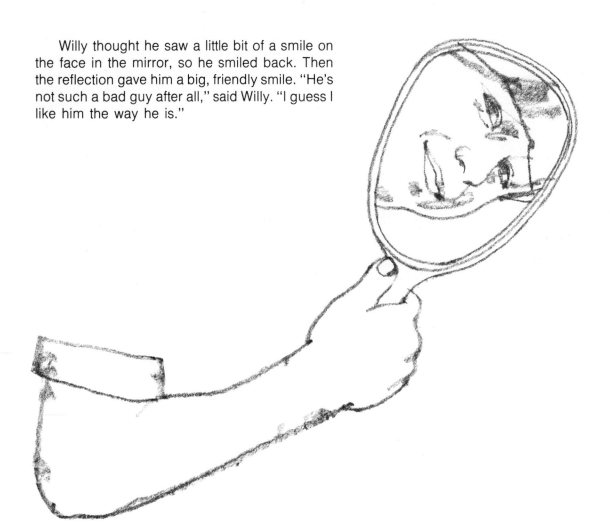

Willy Looks in the Mirror

Problems

1. "I'm thinking of a number that's less than 6 and more than 7," said Ferdie.

What number could it be? (None—but a child might suggest $6\frac{1}{2}$.)
What's wrong with what Ferdie said?

2. "This is the third time today that I've asked you to hang up your coat," said Mark's father.

What will the next time be? (fourth)

3. "How old are you?" somebody asked Ferdie. Ferdie held up a cube. "One year for each side of this cube, and one year besides," he said.

Can anyone figure out how old Ferdie is? (7) [Allow the children to use their response cubes and to take time to work it out.]

4. Mrs. Nosho took 5 children for a ride in her new car. The next day she took 5 more children for a ride.

How many children did she take for a ride altogether? (10)

5. Poor Ferdie dropped a nickel down the drain and couldn't reach it. A big boy with long arms came along and said, "I'll get it for you if you'll pay me 10 cents."

Should Ferdie do it? (no)
Why would it be a silly thing to do?

6. Manolita went fishing with Mrs. Nosho. She caught 1 fish at two o'clock, 1 at three o'clock, and 1 at four o'clock.

How many fish did she catch altogether? (3)

7. Portia can jump 1 meter. Ferdie can jump 10 centimeters farther.

How far can Portia jump? (1 m)

8. When Mrs. Mudanza stretches as far as she can, she can reach 2 meters up.

If she stands on a stool 1 meter high, how high up can she reach? (3 m)

9. Remember, Mrs. Mudanza can reach 2 meters up if she stretches. She wants to paint a window sill that is 4 meters high.

How high a platform will she need to stand on? (2 m)

Mr. Mudanza Builds a Better Tree

8

Mr. Mudanza Builds a Better Tree

Note: The picture that goes with this story is best shown at the end.

One day Mr. Mudanza looked out at the tree in his backyard. "I've changed everything else around this house," he said, "but that tree is still the same. It's a nice tree, but I think I'll change it a little."

He measured the tree and found it was 6 meters tall. He changed it so it was 5 meters tall.

How could he do that? (by cutting off the top, for example)
How much would he have to cut off the top? (1 m)
What would the tree look like afterward?

Mr. Mudanza took a saw and cut off the top part of the tree, so that the tree was a meter shorter. "H'm," he said, "a little flat, but at least it's different." Then he noticed that the tree had 8 branches on one side and 10 branches on the other side. He wanted the tree to have the same number of branches on both sides, so he changed it a little.

How could he make the tree have the same number of branches on both sides? (by cutting 2 branches off one side)

Mr. Mudanza chopped off 1 branch from the side of the tree that had 10 branches. Then he nailed that branch to the side of the tree that had 8 branches.

How many branches are on each side of the tree now? (9)
Is there anything wrong with this way of changing the tree? (The branch will die.)

Mr. Mudanza was happy to find that there were now 9 branches on one side of the tree and 9 branches on the other side, although one branch looked a bit strange. Then he counted 18 apples growing on the tree. "Not bad," he said, "but I'd like it to have 20 apples."

31

What could he do?
How many more apples does he need? (2)

He bought 2 apples at the store and then tied them to the tree with string. "The tree is almost perfect now," he said. "All it needs is a flag at the top." He took a pole about 1 meter long and put it on top of the tree. Then he attached a flag with a big *M* on it.

How high was the tree before he added the flag? (5 m)
How high is the whole thing now? (about 6 m)
Tell all you know about what the tree looks like. Do you think it is really better than it was before?

Note: The picture that goes with this story may be shown and discussed at this time. However, you might want to have the children draw their own pictures of the tree as they imagine it, before they see the illustration.

Mr. Mudanza Builds a Better Tree

Problems

1. Willy had a favorite branch on the tree in his backyard. One day he counted the leaves on the branch and found there were 12. A few days later there were 15.

What season of the year do you think it was? (spring)
Why?

2. "Whenever I lose a tooth," said Willy, "I find a dime under my pillow the next morning. But today I found only 9 cents and a note that said 'You'll get the rest tomorrow.' "

How much money will Willy find under his pillow tomorrow? (1¢)

3. Mr. Mudanza changed his bed a little by putting new legs on it. Now, when he lies on the bed, his feet are higher than his head.

What can you figure out about the legs on the bed? (longer at one end)

4. Mr. Sleeby's house used to be 10 meters high. Now it is 9 meters high.

What could have happened? (It's sinking; the chimney fell off; and so on.)

5. Poor Willy is carrying a heavy load. He was already carrying 3 books, and then Ferdie got him to carry his books too.

How many books is Willy carrying altogether? (Can't tell.)

What do you need to know before you can tell? (how many Ferdie gave him)

6. Willy lives 3 blocks from school. Ferdie lives 1 block farther away from school than Willy does.

How far from school does Ferdie live? (4 blocks)

7. Portia had 6 pennies. She put 1 of them in her pocket, 1 of them fell down a sewer, and 1 of them was 3 years old.

How many pennies does she have left? (5)

8. "Last night I got to stay up till nine o'clock," said Ferdie. "Tonight I have to go to bed by ten."

What could have happened? (*not* something that would make Ferdie have to go to bed earlier)

9. Karen is about 1 meter tall. Her uncle is about twice as tall.

About how many meters tall is Karen's uncle? (2)

10. Manolita and Portia sat on one side of the seesaw, while Virginia sat on the other. They balanced nicely.

Who is the heaviest? (Virginia)
How do you know? (She balances the other 2.)

Mr. Mudanza Builds a Better Tree

Mark Builds a Birdhouse

9

Mark Builds a Birdhouse

Mark built a big, beautiful birdhouse. He made it out of a wooden box. He put a roof on it, painted it green, and set it on a post in the backyard. Some big birds came and looked at it, but they didn't go in. "No wonder," said Mark's mother. "You forgot something important."

What do you think Mark forgot?

"I see what it is," said Mark. "I forgot to make a hole in the front of the birdhouse so the birds can get in!"

Mark went down the street to Mr. Sleeby's store and borrowed a drill from him. With it Mark drilled a hole about 3 centimeters wide in the front of his birdhouse.

How wide is 3 centimeters? Show with your fingers. [Demonstrate the correct width.]

The next day the big birds came and looked at Mark's birdhouse again, but they still didn't go in.

Why not?

"I'm afraid that hole is too small for those big birds," said Mark's brother. "What you need is a hole that's the same size as the birds."

Mark went back to Mr. Sleeby and told him he needed something to cut a bigger hole in his birdhouse. "Exactly how big?" asked Mr. Sleeby.

"I don't know exactly," said Mark. "It should be a sort of bird-size hole."

"Birds come in all different sizes," said Mr. Sleeby. "You'll have to find out how wide the birds are before you'll know how wide a hole to make."

Mark took a ruler and went out and tried to measure the birds.

Do you think that worked? Why not?

The birds were friendly and Mark had no trouble getting close to them; but every time he reached out to put the ruler against one, it flew away.

Mark's friend Portia said, "I have an idea. Why don't you find a picture of that kind of bird in a book and measure the picture? The picture won't fly away."

Does that sound like a good idea? Why not?

Mark looked through a bird book until he found a picture of a bird that looked just like the birds that had come to his birdhouse. He measured the picture with his ruler and found that the bird was 3 centimeters long. "That can't be right," Mark said.

How did Mark know that couldn't be the right size? (If birds are wider than 3 cm, they surely must be longer.)

Mark took the bird book over to Mr. Sleeby's house and showed the picture to him. "Here's the kind of bird it is," said Mark, "but I can't find out what size it is."

"Oh, I know that kind of bird," said Mr. Sleeby. "There are some birds like that building a nest in my house, up in the attic."

"How do they get into your attic?" Mark asked.

"There's a hole in the roof just big enough for them to get through," said Mr. Sleeby. "I was going to fix it someday, but now I guess I won't."

Does that give you an idea?
How could you find out what size hole to make in the birdhouse?

Mark hurried up to Mr. Sleeby's attic and measured the hole in the roof. It was just 7 centimeters wide.

How wide is that? Show with your fingers.
[Demonstrate.]

Mr. Sleeby didn't have a drill 7 centimeters wide, but he had a little saw that would do the job. He and Mark cut a neat hole 7 centimeters wide in the front of Mark's birdhouse. Before long the birds came back. This time they went inside and came out again and soon began bringing grass, string, and feathers to put inside their new home.

Mark Builds a Birdhouse

Problems

1. "I want to measure the inside of this box," said Mark, "but it's too small to get the ruler inside."

Can you think of some ways to measure the inside of the box? (by using string, a paper strip, and so on)

2. Willy has a piggy bank with different slots for pennies and nickels. Nickels will not fit in the slot for pennies.

Do you think a penny will fit in the slot for nickels? (yes)
Why? (A penny is smaller than a nickel.)

3. Mrs. Nosho likes rings, so Mr. Nosho bought her a ring for every finger on each hand, including the thumbs.

How many rings did he buy her? (10)

4. Manolita had 2 dimes. She spent 1 for a red balloon.

How many dimes did she have left? (1)

5. Portia has 9 baseball cards. Willy has 1 more than that.

How many baseball cards does Willy have? (10)

6. Mark walked 5 blocks to school. After school he went home on the bus.

How many blocks did he walk altogether? (5)

7. For dinner Mr. Sleeby ate 2 ears of sweet corn and 2 sausages. He left the corncobs on his plate.

How many corncobs were on the plate? (2)

8. A mother duck had 4 ducklings. When she wanted to go somewhere, she quacked so the ducklings would all follow her. One day when she quacked, 1 duckling came out of the barn and 2 came out from under a bush. After a while another duckling came out of a pile of straw.

How many ducklings were still lost? (zero)

9. Ferdie lives in an apartment on the second floor. One day he looked out his window and said, "I wonder how far it is to the ground."

How could he find out? Can you think of ways to measure how far it is? (Drop a string, then measure it; measure from the first floor to the ground, then double the measure; ask somebody who knows; estimate on the basis of a known height nearby.)

Mark Builds a Birdhouse

Mr. Sleeby Buys a Candy Bar

10

Mr. Sleeby Buys a Candy Bar

One day when Mr. Sleeby was walking down the street he got very hungry. "I guess I forgot to eat breakfast today," he said, "and lunch too." Just then he saw a candy machine with his favorite candy bar in it, a Chewy-Gooey. He looked closely. A label for the Chewy-Gooey bar looked like this [show the illustration or write 10¢ on the board].

How much money is that?

Mr. Sleeby looked through his pockets. He found 6 cents in one pocket, 2 cents in another pocket, and 1 cent in another.

How much money did he find? (9¢)
Is that enough? (no)

He counted and found he had only 9 cents. Then he reached down in his shoe and found another penny.

Is a penny the same as a cent?
How much money does he have now? (10¢)

Mr. Sleeby put his 10 pennies into the machine. The machine went "Clatter-clatter-clink-clink."

Did he get a Chewy-Gooey bar?

No, he got his 10 pennies back. He tried again and the same thing happened.

Just then Mr. Mudanza came along. "What's your trouble, Mr. Sleeby?" he asked.

Mr. Sleeby said, "This machine doesn't work, and I'm starving to death."

Mr. Mudanza looked at the machine. He read a little sign. "It says to use nickels, dimes, or quarters. You're using pennies."

"All I have is pennies," Mr. Sleeby said sadly. "I guess I'll have to go hungry."

Is there anything else Mr. Sleeby can do?

"You can trade your 10 pennies for something that will work in the machine," said Mr. Mudanza.

Could Mr. Sleeby trade his 10 pennies for a quarter? Why not?

"A great idea!" said Mr. Sleeby. "I believe a dime is worth 10 cents. Do you have a dime?"
"No," said Mr. Mudanza. "All I have is nickels."

Should Mr. Sleeby trade his 10 pennies for a nickel? Why not?

"A nickel is worth 5 cents," said Mr. Sleeby. "I'll give you 5 pennies and you give me a nickel. Is that a fair trade?"
"Yes," said Mr. Mudanza, "that's a fair trade." Mr. Sleeby gave him 5 pennies and Mr. Mudanza gave Mr. Sleeby a nickel.
Mr. Sleeby put the nickel in the machine, but it still wouldn't give him a Chewy-Gooey bar. "This machine still doesn't work," he said. "I put in a nickel just as it said, but it didn't give me anything."

Did Mr. Sleeby put enough money into the machine? (no)
How much does a Chewy-Gooey bar cost? (10¢)
How many cents is a nickel? (5)
How many nickels does Mr. Sleeby need? (2)

"A nickel is only 5 cents," said Mr. Mudanza. "A Chewy-Gooey bar costs 10 cents. You have to put in 2 nickels."
"But I have only 1 nickel," said Mr. Sleeby.

How could Mr. Sleeby get another nickel?

"You still have 5 more pennies," said Mr. Mudanza. "If you give them to me, I'll give you another nickel."
Mr. Sleeby gave Mr. Mudanza the 5 pennies and got another nickel. Then Mr. Sleeby looked at the 2 nickels in his hand and his face grew sad. "I used to have 10 pennies and now all I have are these 2 nickels. I think I'm losing money on this deal."

Is Mr. Sleeby losing money? Why not?
How many cents are 2 nickels worth? (10)

Mr. Sleeby put the 2 nickels in the machine and out dropped a Chewy-Gooey bar. Mr. Sleeby opened it and began eating it. "What a marvelous candy bar!" he said. "Why, it's worth more than 2 nickels. It's almost worth a dime!"

Does that make sense? Why not?

Mr. Sleeby Buys a Candy Bar

Problems

1. Manolita had a nickel. She traded it for pennies at Mr. Sleeby's store.

How many pennies did she get? (5)

2. Portia ate half a cupcake. Ferdie ate just as much as Portia.

How many cupcakes did the 2 of them eat together? (1)

3. Ferdie is 1 year older than Portia. "Someday I'll catch up," says Portia.

Will she? Why not? (Ferdie will always be Portia's older brother.)
How old will Portia be when Ferdie is 10? (9)

4. "I have 3 coins in my purse," said Mrs. Nosho, "and together they make 15 cents."

Can anyone figure out what they are? (3 nickels)

"In my coat pocket I have only 2 coins," said Mrs. Nosho, "and together they make 15 cents."

Can anyone figure out what they are? (a dime and a nickel)

"That's nothing," said Ferdie. "I have 4 coins in my pocket, and together they make 8 cents. I'll bet nobody can figure out what I have!"

Can you? (a nickel and 3 pennies)

5. Mr. Mudanza had a desk with 6 empty drawers in it. He filled 2 of the drawers with dirt and is growing mushrooms in them.

What does he have now? (a desk with 4 empty drawers, plus 2 drawers with mushrooms)

6. Loretta the Letter Carrier and her husband, Roger, had 2 dogs and a cat. Yesterday one of the dogs had 3 puppies.

How many dogs do they have now? (5)

7. Loretta and Roger's cat had 5 kittens. Loretta gave 1 to Manolita, 1 to Willy, and 1 to Mark. But Mark's mother wouldn't let him keep his, so he had to give it back.

How many kittens do Loretta and Roger have now? (3)

8. Portia had 9 dandelions. She gave away all but 2 of them.

How many are left? (2)

9. Mr. Sleeby gets lost every time he goes downtown at night. He went downtown 6 times last week.

How many times did he get lost? (Can't tell.)
What do you need to know before you can be sure? (how many times he went at night)

Mr. Sleeby Buys a Candy Bar

Mrs. Nosho's Fish Stories

11

Mrs. Nosho's Fish Stories

Note: The picture that goes with this story is best shown halfway through the reading.

Mrs. Nosho likes to go fishing and she likes to tell other people about it, but sometimes she doesn't tell enough. One day when she got back from a fishing trip, the children were all out in front of her house to meet her.

"How many fish did you catch?" they asked.

"Well," said Mrs. Nosho, "in the morning I caught 4 fish and in the afternoon I caught some more. So you should be able to figure out how many I caught altogether."

Can you figure it out? Why not?

"Wait a minute," said Mark. "How many fish did you catch in the afternoon?"

"I caught only 2 fish in the afternoon," said Mrs. Nosho. "They weren't biting as well then."

Now can you figure out how many fish she caught altogether? (6)

"You caught 6 fish!" the children all shouted, except for Ferdie. Ferdie thought Mrs. Nosho had caught only 2 fish.

What did Ferdie forget about? (the fish she caught in the morning)

"Did you catch any big fish?" asked Willy.

"Indeed I did!" said Mrs. Nosho. "I caught one that is almost as big as the biggest fish I ever caught."

"How long is the fish?" Willy asked.

"You should be able to figure that out for yourself," said Mrs. Nosho, "when I tell you that the big fish I caught today is only 1 centimeter shorter than the biggest fish I ever caught."

Can you figure it out? (no)

"I wish I knew the answer," said Willy.

"Wait a minute," said Mark. "I have a question, Mrs. Nosho."

43

What question do you think Mark is going to ask?

Mark asked, "How long was the biggest fish you ever caught?"

"It was 41 centimeters long," said Mrs. Nosho.

Now can you figure out how long the fish is that Mrs. Nosho caught today? How much shorter is it than the biggest fish? (1 cm)

"You caught a fish 40 centimeters long!" cried Portia.

How long is that? Show with your hands. [Demonstrate.]
Is that very long for a fish? (for some kinds of fish, but not for the biggest fish a person has ever caught)

Mrs. Nosho took a basket out of her car and showed the children the 40-centimeter-long fish.

Note: This is a good point at which to show the picture that goes with the story.

"You may think this is a big fish," Mrs. Nosho said, "but it's nothing compared to the one that got away."
"How long was it?" Ferdie asked.
"I don't know how long it was," said Mrs. Nosho, "but it had eyes as big as my husband's pocket watch."
"Wow!" said Ferdie. "A giant fish!"
"Wait a minute," said Mark. "I have a question."

What question do you think Mark will ask?

"How big is Mr. Nosho's pocket watch?" Mark asked.
"About a centimeter wide," said Mrs. Nosho.

How wide is that? Show with your fingers. [Demonstrate.]

"Yes," said Mrs. Nosho, "my husband has the smallest pocket watch I've ever seen. And the fish didn't have very big eyes, either. But it was heavy. I'll bet that fish weighed twice as much as my Aunt Minnie's hairbrush."
"That's not very heavy," said Manolita. "Everybody knows that a hairbrush doesn't weigh much!"
"Wait a minute," said Mark. "I have another question."

What question do you think Mark will ask this time?

"How much does your Aunt Minnie's hairbrush weigh?" Mark asked.
"It weighs 5 kilograms," said Mrs. Nosho. "It is a very big hairbrush. My Aunt Minnie uses it to brush her pet giraffe."

Now can you figure out how many kilograms the fish weighed? (10)

Mrs. Nosho's Fish Stories

Problems

1. "I always carry 5 pens with me," said Mrs. Nosho, "but I've lost 2 of them. I'm on my way to buy some more."

How many pens does she have with her now? (3)
How many pens do you think she will buy? (2)

2. Manolita is having a birthday party. She just blew out 6 candles on the cake. Two candles are still burning.

How old is Manolita? (8—7 if you include 1 good-luck candle)

3. Portia weighed 20 kilograms. Over vacation she grew 3 centimeters and gained 0 kilograms.

How much does she weigh now? (20 kg)

4. Ferdie had 5 tadpoles. Something happened to 1 of them.

How many tadpoles does he have now? (Can't tell.)
What do you need to know? (whether that 1 is no longer a tadpole)
[If a child insists that 1 turned into a frog or died, ask]:
Do you know for sure? What if it just got bigger?
How many would he have then?

5. "I'll bet my paper airplane can fly farther than yours," said Manolita. Willy threw his airplane and it went 4 meters. Manolita threw hers and it went 3 meters. Then she threw it again and it went 3 more meters.

Whose airplane can fly farther? (Willy's)

6. Mark is growing fast. He grew 5 centimeters in just 9 months.

How tall is he now? (Can't tell.)
What do you need to know before you can figure it out? (how tall he was)

7. Ferdie's pencil broke near the middle. He wonders if he has more than half or less than half of it left to write with.

How could he find out? (by seeing which piece is longer)

8. Manolita walked to her friend Ivan's house and then walked home. Ivan lives 2 kilometers away.

How far did Manolita walk altogether? (4 km)

9. Willy can wear Ferdie's shoes, but Ferdie can't get into Willy's shoes.

Who has bigger feet? (Ferdie)

Mrs. Nosho's Fish Stories

Manolita's Magic Number Machine

12

Manolita's Magic Number Machine

Ever since Manolita saw the giant computer her mother works with, she often dreams about magic machines. One night in a dream she came upon a big machine with an opening on the top and a door at the bottom. Manolita had no idea what kind of machine it was, but she thought she might find out by putting some money in it. She put a penny into the top of the machine. The machine went "Glinka-Glinka" and out popped 3 pennies at the bottom.

Then Manolita put in a pencil. The machine went "Glinka-Glinka" again and out popped 3 pencils! "This is great!" chirped Manolita. "I wonder what will happen if I put in 3 pencils."

What do you think will happen? [Offer praise to children who say 5 or 9 pencils will come out.]

Manolita dropped in the 3 pencils. The machine went "Glinka-Glinka" and out came 5 pencils. Manolita had thought there might be a few more than that, but she was happy to have 5 pencils anyway. "I have a great idea!" she said. "I'll put all my money in the machine, and maybe it will give me back so much money that I'll be rich."

She opened her piggy bank and took out all her money. There were 30 pennies. Eagerly she put all 30 pennies into the top of the machine.

How many pennies do you think she will get back? (32)
Would it have been better for Manolita to put in the pennies one at a time?

The machine went "Glinka-Glinka" and out came 32 pennies. Manolita thought the machine might have made a mistake. She put the 32 pennies in, the machine went you-know-what again, and out came 34 pennies. Manolita tried it once more. She put in the 34 pennies, and this time the machine gave back 36 pennies. "I think I have it figured out," said Manolita, as she put in the 36 pennies.

How many pennies will she get back this time? (38)

The machine went "Glinka-Glinka" and out came a heap of pennies. Manolita counted them. "Just as I thought," she said. "When I put in 36 pennies, the machine gave me back 38. Every time I put some things in the machine I get more."

How many more? (2)

"I always get 2 more than I put in," said Manolita. "So, if I put my 2 turtles in the machine ..."

What will happen? (She'll get 4 turtles.)

Manolita did it; and, sure enough, the machine gave back 4 turtles. Then Manolita took 5 of her mother's silver spoons and put them in the machine.

How many silver spoons will she get back? (7)

The machine gave back 7 silver spoons, 2 more than Manolita had put in. Then Manolita had a wild idea: "What if I climb into the machine myself?"

What do you think will happen?

We'll never know for sure, because Manolita woke up just as she was starting to climb into the Magic Number Machine.

Manolita's Magic Number Machine

Problems

1. Another night Manolita dreamed about a magic money-changing machine. If you put in a dime, it gave back 11 pennies. If you put in a quarter, it gave back 26 pennies.

How many pennies would it give back if you put in a nickel? (6)

2. Manolita put 10 potatoes into another magic machine. She got back 8 potatoes. She put in 5 books and got back 3.

What was the machine doing? (taking away 2)

3. Willy had 6 marbles. He lost half of them.

How many did he lose? (3)

4. Ferdie had 10 cents. He bought 2 pieces of candy.

How much money did he have left? (Can't tell.)
What do you need to know before you can tell? (the cost of the candy)

5. "That tree is 2 years older than I am," said Mr. Burns, "and I am 40 years old."

How old is the tree? (42)

6. Portia lives on the second floor of an apartment building. Her friend Janet lives on the fourth floor.

How many floors up does Portia have to go to visit Janet? (2)
How many floors up does Janet have to go to visit Portia? (She doesn't go up; she goes down.)

7. Willy had a frog that could jump very high. One day it was hopping across the lawn. Suddenly it jumped 2 meters straight up. Then it fell back down to the lawn.

How far did it fall? (2 m)

8. Mark invited 5 boys to his party. They all came.

How many boys were at the party altogether? (6)

9. Ferdie is usually 10 centimeters taller than Portia, but now Portia is standing on a box that is 10 centimeters high. They are both standing against the wall.

Who comes up higher on the wall—Ferdie or Portia? (Neither; they're the same.)
How do you know?

10. Mrs. Nosho counted 22 roses on her rose bush in the morning. In the afternoon there were 18.

What could have happened? (Some fell off; some were picked; and so on.)

11. Mrs. Mudanza had a mirror with 4 light bulbs on one side and 4 on the other. One day Mr. Mudanza changed it a little. He took out 2 of the light bulbs and put in a toothbrush drier instead.

What kind of mirror does Mrs. Mudanza have now?
How many light bulbs does it have? (6)

Manolita's Magic Number Machine

Mr. Mudanza Makes Lunch

13

Mr. Mudanza Makes Lunch

One day it was Mr. Mudanza's turn to make lunch. Even when he made lunch he couldn't help changing everything a little.

The first thing Mr. Mudanza did was take out 3 stalks of celery and put them on a plate. But that didn't look like very much, so he changed the celery a little. He cut each stalk of celery in half.

How many pieces of celery are there now? (6)
Is there more celery than before? (no, just smaller pieces)

Next Mr. Mudanza took out 6 cupcakes. "These look like tasty cupcakes," said Mr. Mudanza, "but they're on the small side. I think I'll change them a little."

He took 2 cupcakes, put one on top of the other, and squeezed them until they were mashed together.

What did he make? (1 big, messy cupcake)

Then he did the same thing with 2 other cupcakes, and then with 2 other cupcakes after that.

Now how many cupcakes does Mr. Mudanza have altogether? (3)

"Three big cupcakes," said Mr. Mudanza. "That's just right: 1 for me, 1 for my dear wife, and 1 for my little Manolita."

Next Mr. Mudanza took 2 pitchers of juice out of the refrigerator. There were 3 cups of grape juice and 2 cups of tomato juice. He mixed them together.

How many cups of juice does he have now? (5)
What do you think it tastes like?

Finally Mr. Mudanza took out some vegetables. There were green peas and chickpeas all mixed together in a bowl. He measured them with a big spoon. There were 10 spoonfuls. "We have 10 tasty spoonfuls of vegetables," said Mr. Mudanza, "but I think I'll change them a little."

He carefully picked out all the green peas. When he was done, there were 6 spoonfuls of chickpeas left.

Can you figure out how many spoonfuls of green peas he took out? (4)

At lunchtime Manolita was amazed. She thought her father had done some magic. "These cupcakes are much bigger than they used to be," Manolita said. "What kind of magic did you use on them, Papa?"

"Just a little squeeze-power," said Mr. Mudanza.

Then Manolita tasted the juice. She took a sip, made a face, and put it down. "I don't know what kind of magic you used on this juice," said Manolita, "but I wish you'd use some more magic and change it back to the way it was!"

Could Mr. Mudanza change the juice back to the way it was? Why not?

Mr. Mudanza Makes Lunch

Problems

1. Mr. Sleeby said, "I started out with 3 tickets to the circus, and then either I gave 1 away or somebody gave me 1—I can't remember which."

How many tickets does Mr. Sleeby have now? (2 or 4—can't tell.)
What do you need to know to be sure?

2. When his father came into the house, Willy was standing in the middle of the stairs. He went up 2 steps. Then he went down 3 steps and stayed there.

Is Willy higher or lower now than when his father came in? (lower)
How much lower? (1 step)

3. Mr. Sleeby has 3 lamps in his living room. Each lamp has a light bulb in it, but 2 of the bulbs have been burned out for a long time, so in the evening Mr. Sleeby always sits by the third lamp. Last night the bulb in that lamp burned out, too, so Mr. Sleeby had to go to bed. This morning Mr. Sleeby went to the hardware store and bought 4 light bulbs.

Will Mr. Sleeby have enough bulbs for all the lamps in his living room? (yes, and 1 extra)

4. Mrs. Nosho bought 2 skirts and a sweater for her daughter Patty. Then she bought exactly the same things for her other daughter, Pitty.

How many skirts did she buy altogether? (4)
How many sweaters did she buy altogether? (2)

5. Mr. Mudanza had a belt with 8 holes in it, but he changed it a little. He filled in half the holes.

How many holes are left in the belt? (4)

6. Portia has an extra wheel fastened to the back of her wagon in case she has wheel trouble.

How many wheels are on the wagon? (5)

7. Ha-ping used to walk a block to school. Now she has to walk twice as far.

How far does she have to walk now? (2 blocks)

8. Chewies cost 2 cents apiece. Ferdie bought 1 of them with money from his piggy bank. Mark bought 3 of them with money his father gave him.

How much did Ferdie pay? (2¢)

9. "I'll give you 10 cents, Mark, if you mail this letter for me," said Mr. Breezy, "and I'll give you twice as much if you go very fast." Mark ran like an antelope (that's very fast).

How much money did Mark get? (20¢)

Mr. Mudanza Makes Lunch

Manolita's Magic Minus Machine

14

Manolita's Magic Minus Machine

Manolita was dreaming about a magic machine again. This machine was just like the ones she had dreamed about before, except that if you put in 5 things, it would give back 4; if you put in 6 things, it would give back 5; if you put in 9 things, it would give back 8.

What was the machine doing? (taking away 1)

When Manolita woke up, she decided that she could build a magic machine like that. She found a big, big box. On it she painted a sign that said "MAGIC MINUS MACHINE. Whatever you put in the top, you get back 1 less at the bottom. FREE!"

Manolita put the magic machine out by the sidewalk and hid inside it. Soon children began flocking around the machine and reading the sign. "It's free!" Portia said.

Mark was the first to try it. He put in 7 sticks. The machine went "Glinka-Glinka" and out came 6 sticks at the bottom.

What do you think made the machine go "Glinka-Glinka"?

What do you think happened to the other stick that Mark put into the machine?

Ferdie wanted to try the machine next. He put in 9 marbles.

How many marbles will he get back? (8)

"Hey," said Ferdie, "the machine kept 1 of my marbles!" Ferdie was angry and walked away.

Mark tried the machine again. He put in 4 candies.

How many candies will he get back? (3)

The machine went "Glinka-Glinka" and out came 3 candies at the bottom. Mark didn't like that very much, but he said, "I'm going to put them back in the machine, and maybe this time more will come out." He put his 3 candies into the machine.

Will he get back more than 3? (no)

55

This time the machine gave him back only 2 candies. That made Mark angry, and he walked away.

Portia felt in her pocket and found 5 pumpkin seeds that she had been saving to plant. She dropped them into the Magic Minus Machine and waited eagerly to see what would happen.

What will happen? (She'll get back 4 seeds.)

The machine went "Glinka-Glinka" and out came 4 pumpkin seeds. "Nasty machine!" said Portia, and she walked away. Soon none of the children would have anything to do with Manolita's Magic Minus Machine.

Why not?

Then Willy the Wisher came along. He had just finished eating a banana. "I wish I had someplace to put this banana peel," said Willy. "I wish there was a wastebasket right here." Then he noticed the Magic Minus Machine. He put the banana peel into the top.

What do you think will come out the bottom? When you put in 1 banana peel, how many banana peels do you get back? (zero)

The machine went "Glinka-Glinka," but no banana peel came out the bottom. After that, whenever people had some trash to get rid of, they put it into Manolita's Magic Minus Machine.

Manolita's Magic Minus Machine

Problems

1. Manolita dreamed about another magic machine. She put in 3 acorns and got back 6. She put in 2 cards and got back 4. She put in 1 card and got back 2.

What was the machine doing? (doubling the number)
What will Manolita get back if she puts in 4 sticks? (8 sticks)

2. All the children were eating pizza. Willy was the second one to finish. Manolita was the fourth.

Who ate faster—Manolita or Willy? (Willy)
How can you tell?

3. Portia saves 2 pennies every day.

How long will it take her to save 10 cents? (5 days)

   Ferdie saves only 1 penny a day.

How long will it take him to save 10 cents? (10 days)

4. Portia was learning to walk on stilts. The first day she could take 2 steps before falling. The next day she could take 4 steps. The next day she could take 6 steps.

How many steps would you guess she could take the day after that? Why do you think so? (Any answer will do, although 8 is the obvious one. What is of interest is the reason given.)

5. Willy got 4 gumdrops. "I'm only going to eat 2 gumdrops every day," said Willy, "so they'll last a long time."

How many days will they last? (2)

6. Portia walked 1 block with her friend Taro. Then she walked 3 times as far by herself.

How many blocks did Portia walk altogether? (4)

7. Willy walked half a block to the library. Then he walked half a block to get home.

How many blocks did he walk altogether? (1)

8. Ferdie decided he was going to be nice to 5 people today. He tried and tried, but so far he has managed to be nice to only 2 people—himself and Mr. Sleeby.

How many more people does Ferdie need to be nice to? (3)

9. "This is the fourth time I've been to the zoo," said Portia.

How many times had she been to the zoo before? (3)

10. Portia had 4 dolls. She gave a doll to Willy and a tennis ball to Manolita.

How many dolls does she have left? (3)

Manolita's Magic Minus Machine

How Mrs. Nosho Doubled Her Money

15

How Mrs. Nosho Doubled Her Money

"You have 2 of everything, Mrs. Nosho," said Mark. "You must be rich."

"That's right," said Mrs. Nosho.

"How did you get that way?"

"Simple," said Mrs. Nosho. "By getting lots of money."

"But how did you do that?" Portia asked.

"I'll tell you how it all started," said Mrs. Nosho. "One day when I was a young girl I found a quarter on the sidewalk. I used it to buy something."

"What?" Ferdie asked.

"A handkerchief, of course."

"Did you have a cold?"

"No," said Mrs. Nosho, "but I soon met a woman who did, and she gave me some money for the handkerchief. That's how I started to get rich."

"Wait a minute," said Mark. "I have a question."

What question do you think Mark will ask?

"My question," said Mark, "is how much did you sell the handkerchief for?"

"Twice as much as I paid for it," said Mrs. Nosho.

How much did Mrs. Nosho pay for the handkerchief? (1 quarter)
So how many quarters did she get for the handkerchief? (2)
How much money is 2 quarters? (50¢)

"So you had 50 cents," said Mark.

"Yes," said Mrs. Nosho. "And with the 50 cents I bought some doughnuts. Then I sold the doughnuts to my friends, you see, and made more money. Can you figure out how much I made?"

"A nickel?" Ferdie guessed.

"Wait a minute," said Mark.

What questions does Mark need to ask?

"How many doughnuts did you have and how much did you sell them for?" Mark asked.

"I had 10 doughnuts," said Mrs. Nosho, "and I sold them for 10 cents apiece."

Can you figure out how much money she got altogether from selling the doughnuts? ($1) Try counting by tens.

"You got a dollar!" Portia piped up.

"That's not so much," said Ferdie. "I had a dollar once."

"Ah," said Mrs. Nosho, "but I used my dollar to buy a sick canary."

"Why did you buy a sick one?" Portia asked.

"So that when it got well and started to sing I could sell it for more money. So I sold it for more than I paid for it, and . . ."

"Wait a minute," said Mark.

What is Mark going to ask?

"How much did you sell the canary for?" Mark asked.

"For a dollar more than I paid for it," Mrs. Nosho answered.

How much money did she have after she sold the canary? ($2)

"You had 2 dollars!" the children said.

"Right," said Mrs. Nosho. "And I used the 2 dollars to buy a shoe."

"One shoe?" asked Ferdie.

"That's right."

"But one shoe isn't worth anything," Ferdie said.

"This one was," said Mrs. Nosho, "because I knew a man who had one shoe just like it for the other foot. So he was glad to pay me what I asked for the shoe, and with that money . . ."

"Wait a minute," said Mark.

What question is Mark going to ask?

"How much money did you ask the man to pay for the shoe?" Mark asked.

"Why, I asked him to pay twice as much as I paid," said Mrs. Nosho. "I always like to double my money."

How much did Mrs. Nosho pay for the shoe? ($2)

What's twice as much? So how much money did she get for the shoe? ($4)

"Four dollars!" said Ferdie. "You really were getting rich!"

"But I didn't stop there," said Mrs. Nosho. "I used the 4 dollars to buy a clock with no hands."

"What good is a clock with no hands?" Portia asked.

"I don't know," said Mrs. Nosho. "In fact, I never could find anyone who wanted to buy the clock, so I still have it. If you happen to know anyone who wants a clock with no hands, I'll sell it for a good price."

"What price?" asked Mark.

"Why, only twice as much as I paid for it," said Mrs. Nosho.

How much did she pay for the clock? ($4)
How much does she want to sell it for? ($8)

"But wait," said Manolita, who had been quiet until then. "You were going to tell us how you got rich. If you bought the clock for 4 dollars and you still have the clock, then you couldn't get rich that way."

Why not? How much money did Mrs. Nosho have left after she bought the clock with no hands? (none)

"Oh, I forgot to tell you one little thing," said Mrs. Nosho. "When I opened up that old clock with no hands, I found it had half a million dollars inside."

How Mrs. Nosho Doubled Her Money

How Mrs. Nosho Doubled Her Money

Problems

1. If Mr. Breezy works all day, he can wash half the windows in his house.

How many days will it take him to wash all the windows? (2)

2. Mark had 8 shells that he found on the beach. He showed 2 of them to his friend Willy.

How many shells did Mark have then? (8)

3. "What a smart baby!" said Portia. "He spoke his first word when he was only 10 months old, and he spoke his second word when he was only 9 months old."

What's wrong with what Portia said? (He spoke the second word before the first.)

4. A friendly baker gave the children 2 Danish rolls. "Break each one in half," he said.

How many pieces did the children have after they broke the rolls? (4)

   There were 8 children.

Did each of them get half a roll? (no)
What could they do? (break each of the 4 pieces in half)

5. Ferdie had 8 marbles and Mark had 4. Mr. Breezy gave 2 marbles to the boy who had less.

How many marbles does Ferdie have now? (8)

6. The Mudanzas used to be able to bake enchiladas in 15 minutes, but Mr. Mudanza changed the oven a little. Now it takes 10 minutes longer.

How many minutes does it take to bake enchiladas now? (25)

7. Once Mr. Nosho grew a beard. It was 20 centimeters long, but Mr. Mudanza changed it a little. He cut 18 centimeters off.

**How long was Mr. Nosho's beard then? (2 cm)
How long is that? Show with your fingers.** [Demonstrate.]

8. Mr. Mudanza had a necktie that was 84 centimeters long. He cut a centimeter off one end and a centimeter off the other end.

How long is the necktie now? (82 cm)

9. There is a snail in Willy's fishbowl that is trying to crawl up the side. Every day it climbs up 5 centimeters and every night it slides down 5 centimeters.

How far up will the snail get after 4 days and 4 nights? (0 cm)

10. There is another snail in Willy's fishbowl that does a little better. Everyday it climbs up 5 centimeters, but at night it only slides down 3 centimeters.

How Mrs. Nosho Doubled Her Money

How far up will the snail get after 4 days and 4 nights? (8 cm) [Suggest that the children use a ruler or number line to work it out.]

11. Mr. Mudanza used to be able to get 5 channels on his TV set. He made the antenna higher, and now he can get 3 more channels.

How many channels can he get now? (8)

12. Manolita had 7 marking pens. She sold some of them to Portia for 5 cents apiece.

How many marking pens does Manolita have left? (Can't tell.)
What do you need to know before you can tell? (how many she sold)

How Mrs. Nosho Doubled Her Money

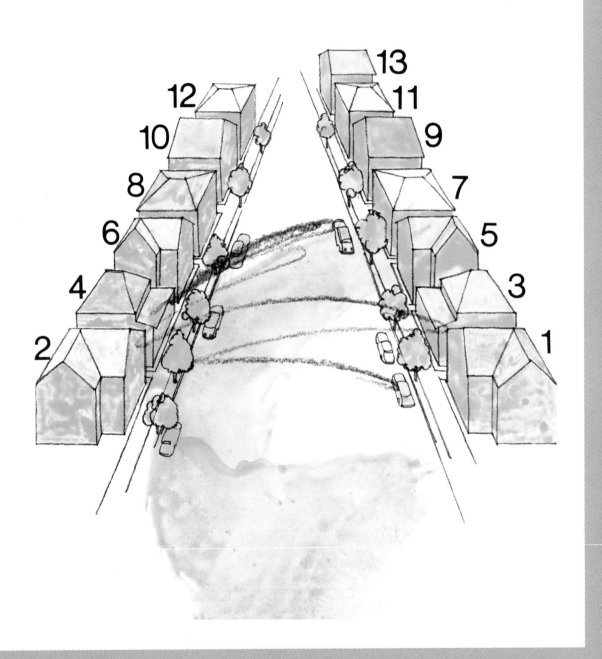

The Third House on Fungo Street

16

The Third House on Fungo Street

One day Ferdie and Portia were walking along with their friend Loretta as she was delivering the mail. "Here's a hard one to figure out," said Loretta, looking at an envelope. "This is a letter for someone named Sandy Bright, and the only address on it is 'Third House on Fungo Street.'"

"That should be easy," said Ferdie. "Fungo Street is so short there aren't very many houses on it."

"Then perhaps you can tell me which house it is," said Loretta the Letter Carrier, as they started walking along Fungo Street.

There are just 13 houses on Fungo Street, and this map shows where they are. The numbers are house numbers. [Show the illustration.]

"I know which house it is," said Ferdie. He counted off "1, 2, 3" and pointed to house number 6. "That's the third house on Fungo Street," he said.

Ferdie marched up to the door of house number 6 and called, "We have a letter here for Sandy Bright!"

"Nobody named Sandy Bright lives here," said a gruff man who came to the door.

What could be wrong? What other house could be the third house on Fungo Street? (number 5)

"I know," said Ferdie. "It must be house number 5, across the street. That's the third house on Fungo Street too, only on the other side of the street!"

They went across to house number 5, but no Sandy Bright lived there either. "I give up," said Ferdie. "Whoever wrote that letter didn't know where Sandy Bright lives."

Then Portia said, "I have an idea where the third house on Fungo Street might be."

Do you have an idea where it could be?

"Maybe it's the third house from the *other* end," said Portia.

"We'll try your idea," said Loretta.

65

Which house is the third house from the other end? (number 9)
Which other house is also the third one from the other end? (number 8)

They started at the other end and counted house number 13, house number 11, and house number 9. "This must be it," said Ferdie. "House number 9 is the third house on Fungo Street."

So they knocked on the door. Mr. Sleeby came to the door and said hello. Loretta the Letter Carrier said, "You don't have anyone named Sandy Bright living in your house, do you, Mr. Sleeby?"

"Not that I can think of," said Mr. Sleeby.

"There's only one other house it could be," said Portia.

What house is that? (number 8)

They went up to house number 8. A big man with reddish-brown hair was sitting on the front porch, smoking a pipe. "Is there anyone here named Sandy Bright?" asked Loretta the Letter Carrier.

"I'm Sandy Bright," said the man. "I wondered what you were doing, walking up and down the street that way."

They gave him the letter and walked away. "I'm afraid I have another hard letter to figure out," said Loretta. "This one is just addressed to 'Otto, The Fifth House on Fungo Street.'"

Which houses could be the fifth house on Fungo Street?
Can you find 4 different ones? (numbers 5, 4, 10, 9)

First they tried house number 5, but no one named Otto lived there. Then they tried house number 4. No Otto. Then they started counting from the other end and tried house number 10. Still no Otto.

Which house haven't they tried yet? (number 9)
Who lives there? (Mr. Sleeby)

"It must be house number 9," said Portia. "But that's Mr. Sleeby's house."

"Hey," said Ferdie, "something's wrong here. Mr. Sleeby's house was the third house on Fungo Street. How can it be the fifth house on Fungo Street too?"

Can you figure out how it can be the third house and the fifth house at the same time?

Ferdie thought for a minute and then said, "I get it. Mr. Sleeby's house is the third house from one end and it's the fifth house from the other end."

They went up to Mr. Sleeby's house again and knocked on the door. "Hello again, Mr. Sleeby," said Loretta. "This time we're looking for somebody named Otto, who lives in the fifth house on Fungo Street."

Mr. Sleeby was delighted. "That's me," he said, "Otto Sleeby. Ah, I see you have my letter. I wrote it myself."

Portia asked, "Why did you write yourself a letter, Mr. Sleeby?"

"I can't remember," said Mr. Sleeby. "I'll have to read it and find out. Maybe it contains important news."

"If you write an answer to that letter," said Loretta, "I hope you'll put your whole name and your house number on it. Otherwise you may never get it."

"That would be dreadful," said Mr. Sleeby. "Then I'd never know what happened, would I?"

The Third House on Fungo Street

The Third House on Fungo Street

Problems

1. Ferdie sat in the first chair. Portia sat in the fourth chair.

How many chairs are between them? (2)
Which chairs are they? (second and third)

2. Mr. Sleeby decided to give a prize to the third person who came into his toy store that day. First came Mark; then came Manolita; then Loretta the Letter Carrier brought in some mail; then came Portia; and then came Ferdie.

Who got the prize? (Loretta)

The next time Mr. Sleeby decided to give the prize to the fourth *child* who came in. First came Ferdie, then Mark, then Portia, then Loretta the Letter Carrier, then Mrs. Nosho, then Willy, then Manolita, then Ferdie again.

Who got the prize this time? (Willy)

3. Mr. Sleeby drinks half a container of milk a day. He just opened a container of milk this morning.

When will he have to open another container? (in 2 days, or the day after tomorrow)

4. "I'm trying to save up 10 cereal box tops to get a free kite," said Mark. "I've already saved 2."

How many more box tops does he need? (8)

5. Ferdie had 15 cents. His mother gave him a dime to go and buy a newspaper for her.

How much money will Ferdie have after he buys the newspaper? (15¢, if the newspaper costs 10¢)

6. "How many years have you had this toy store?" somebody asked Mr. Sleeby.
 "I don't remember," said Mr. Sleeby, "but I do know I was 40 years old when I got it and I'm 45 years old now."

Can you figure out how long he's had the toy store? (5 years)

7. Manolita's house is a block away from Portia and Ferdie's. Yesterday Manolita walked over to Portia and Ferdie's to play. When it was time for dinner Manolita walked home. Then she walked back to Portia and Ferdie's house to spend the night.

How far did Manolita walk altogether? (3 blocks)

8. All the children are sitting in a row at the movie. Listen and figure out who is sitting next to Willy: Willy is sitting in the third seat. Ferdie is in the first seat, Mark is in the fourth seat, Manolita is in the fifth seat, and Portia is in the second seat.

Who is sitting next to Willy? Who else? (Portia and Mark)

9. Manolita found half a doughnut in the cookie jar, half a doughnut in the refrigerator, and half a doughnut in a paper bag.

If she put them all together, how many doughnuts would she have? (3 halves, or $1\frac{1}{2}$)

10. Mark invited 2 boys for lunch. Each of the boys took his little brother along too.

How many boys went to Mark's house for lunch? (4)

11. Mr. Mudanza had a candle that was 20 centimeters long. He let it burn until only 8 centimeters were left, and then he cut a centimeter off the bottom.

How long is the candle now? (7 cm)

12. Six children had a race down to the beach. Manolita was the fourth child to get there.

How many children got there after her? (2)

The Third House on Fungo Street

The Lemonade War

17
The Lemonade War

One day when Ferdie was walking along the sidewalk he saw Mark standing behind a box. On the box were a pitcher and some glasses and a sign that said Lemonade 5¢.

"What are you doing?" Ferdie asked.
"Selling lemonade," said Mark.
"Do you get to keep the money?"
"Yes," said Mark, "and I've already sold 2 glasses of lemonade."

How much money has Mark made? (10¢)

"That's a great idea," said Ferdie. "I think I'll do it too." So he got some lemonade and glasses and a box from his mother and set up a lemonade stand on the sidewalk, right next to Mark's. But Ferdie was a little greedy. He wanted to make more money than Mark, so he wrote on his sign Lemonade 6¢.

Two children, Janet and Ken, came along. They read both signs and then they bought some lemonade from Mark.

Why do you think they did that?

"Hey!" said Ferdie. "Why didn't you buy from me?"

"You charge too much," Ken said. "Why should we pay you 6 cents when Mark sells it for 5 cents?"

Ferdie thought about it and then he had an idea. He changed the sign so that now it said Lemonade 4¢.

Why did he do that?

Soon Manolita came along. She read both signs and went to Ferdie's stand. She held up a nickel and said, "One glass of lemonade, please."

How much is a nickel worth? (5¢)

"My lemonade is 4 cents," said Ferdie. "Don't you have 4 cents?"
"No."

71

"Too bad," he said. "Come back when you do."

Instead, Manolita went to Mark's stand and bought lemonade from him for a nickel.

What should Ferdie have done? (given change)
How much money should he have given Manolita? (1¢)

"You should have taken the nickel," said Mark, "and given Manolita a penny change."

"Oh," said Ferdie. "Thanks, I'll remember to do that next time."

The next person to come along, Mr. Burns, had a dime. He handed it to Ferdie and asked for a glass of lemonade. Ferdie poured him a glass, took the dime, and gave him back 1 penny.

Was that right? How much should Ferdie have given him? (6¢)
Why do you suppose Ferdie gave him just a penny? (That's what he should have given Manolita.)

Mr. Burns was angry. "That's not enough change," he said. "Here, keep your lemonade and give me back my dime." Then he took his dime to Mark's stand and bought a glass of lemonade with it. Mark gave him the right change.

How much change did Mark give him? (5¢)

"Next time I'll give the right change," said Ferdie.

Along came Mrs. Downey and Mrs. Kamato. They went to Ferdie's stand and each one asked for a glass of lemonade. Ferdie was delighted. "You came to the right place, ladies. Best lemonade. Best prices. And I always give the right change."

Mrs. Downey gave Ferdie a dime. Ferdie carefully counted out 6 cents change for her.

Why did he give her 6 cents?

"Oh," said Mrs. Downey, "but I wanted to pay for *both* glasses of lemonade with my dime."

"Excuse me," said Ferdie, and he gave her 6 more cents change.

Was that the right thing to do? (no)
How much change did Ferdie give Mrs. Downey altogether? (12¢)
How can you tell that's too much? How much money did she give him? (10¢)
Should you ever get more change than the amount you paid? (no)

When Mrs. Downey and Mrs. Kamato left, Ferdie said, "At last I've sold some lemonade. I'm rich! I'm rich! I'm rich!" Then he counted out his money and found that he had less than when he started.

Why did he have less? (He gave too much change.)

"I've been robbed!" Ferdie screamed.

"No, you haven't," said a voice. It was Mrs. Downey, who had come back. "I just wanted to see if you would figure out that you made a mistake," she said. She gave Ferdie 10 cents and told him to be more careful with his money next time.

"There isn't going to be a next time," said Ferdie. "I'm getting out of this business while I still have some money left."

Ferdie picked up his lemonade and glasses and box and sign, and left. As he was walking away he heard Mark shouting, "Get your lemonade here, folks! Only 6 cents!"

What had Mark done? (raised his price)
Why could Mark charge 6 cents now? (no competition)

The Lemonade War

The Lemonade War

Problems

1. "Look at the nickel I found," said Willy.
"I could use a nickel like that," said Ferdie. "I'll give you 4 cents for it."

Should Willy sell it to him? Why not?

2. Mark brushes his teeth 4 minutes almost every day, but yesterday he was lazy and brushed them only half that long.

How many minutes did he brush his teeth yesterday? (2)

3. Manolita bought a toy airplane for 2 dollars and sold it for 1 dollar.

How much money did she make? (none)
How much did she lose? ($1)

4. Mark and Manolita were going fishing with Mrs. Nosho, but first they all had to dig worms. Mrs. Nosho dug 3 worms. Mark dug 1, and Manolita dug 2.

How many worms did the children dig altogether? (3)

5. One week Portia ate a peanut-butter sandwich on Monday and another one on Tuesday. The next week she did the same thing.

How many did she eat altogether? (4)

6. Mr. Sleeby is filling a barrel with water. He has already put in 10 liters.

How many more liters of water does he have to put in to fill up the barrel all the way? (Can't tell.)
What do you need to know? (how many liters the barrel holds)

7. "Be sure to bring back the change," said Portia's mother. Portia went to the store and bought a loaf of bread. She gave the grocer 5 dimes and got back 1 penny.

How much did the bread cost? (49¢)

8. Mr. Mudanza had a 12-string guitar but he changed it a little. He put 4 more strings on it.

What kind of guitar does he have now? How many strings? (16)

9. Manolita had 4 crackers. She broke them all in two and ate them.

How many crackers does she have left? (zero)

10. Mr. Breezy is painting a shed. It has 4 sides. He can paint 2 sides in 1 hour.

How many hours will it take him to paint all 4 sides? (2)

11. Every time Portia dips her paintbrush she can paint 1 side of a board. She wants to paint 3 boards on both sides.

How many times will she have to dip her brush? (6)

The Lemonade War

12. Willy was able to stand on one foot for 8 minutes. "I'll bet I can stand on one foot for 9 minutes," said Ferdie. If Ferdie can last 3 more minutes, he will make it.

How many minutes has Ferdie been standing on one foot so far? (6)

13. Ferdie and Manolita were arguing about whose jacket pocket was bigger. Ferdie put 15 acorns into his pocket. Manolita put 14 acorns into her pocket, and 2 of them fell out.

Whose pocket is bigger? (Ferdie's)
How do you know?

The Lemonade War

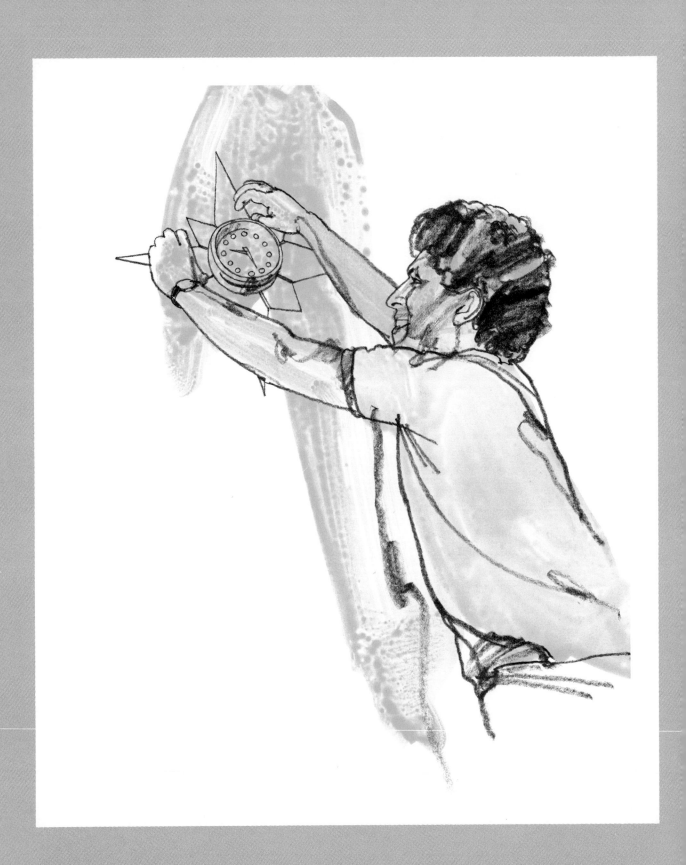

Mr. Mudanza Changes Houses

18

Mr. Mudanza Changes Houses

One day Mr. Mudanza made the mistake of changing his house into a bowling alley, and then his family had no place to live. "You can stay in our summer house," said Mrs. Nosho, who had 2 of everything, even houses. "Just make yourselves at home there and change anything the way you want to."

The Mudanzas were very pleased with the Noshos' summer house; but as soon as they moved in, Mr. Mudanza started to change things a little. There was a clock on the wall that looked like this [show this illustration]:

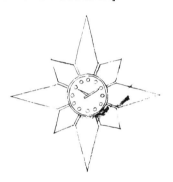

How many points does the clock have? (8)

Mr. Mudanza changed the clock a little by taking off all the long points.

How many points are left? (4)

"I hope the Noshos like a 4-pointed clock," said Manolita.

There was a poster on the wall that looked like this [show this illustration]:

How many corners does the poster have? Count them. (4)

Mr. Mudanza cut off 1 of the corners, so that the poster looked like this [show this illustration]:

How many corners does the poster have now? Count them. (5)

Mr. Mudanza liked the way the poster looked so much that he clipped off the other 3 corners the same way.

How many corners does the poster have now? Get a picture in your mind. (8)

"I hope the Noshos like an 8-cornered poster," said Mrs. Mudanza.

On the floor of the summer house was a very thick rug. It was 4 centimeters thick.

How thick is that? Show with your fingers. [Demonstrate.]

Mr. Mudanza changed the rug a little. He cut it in half and put one half on top of the other.

How thick is it now? (8 cm)

"I'm sure the Noshos will be happy to have a rug 8 centimeters thick on the floor," said Mr. Mudanza.

"I'm not so sure," said Mrs. Mudanza. "The rug is twice as thick as it was before, but it isn't as long."

How long is it now? (half as long)

There was a fishbowl in the summer house. Mr. Mudanza changed the bowl by cutting it in half from top to bottom.

How many fishbowls do the Mudanzas have now? (none)

They don't have any fishbowl, because neither half holds water. So they put the fish in the bathtub. Now they have no fishbowl and no place to take a bath, either.

Mr. Mudanza Changes Houses

Problems

1. Do you know how a hairpin is shaped? Like this [show the illustration or, if you're working with a large group, draw a hairpin on the chalkboard]:

Mrs. Mudanza had a hairpin that was 5 centimeters long. Mr. Mudanza changed it a little. He straightened it so there is no bend any more.

How long is the hairpin now? (about 10 cm)

2. Mr. Mudanza had a scarf 80 centimeters long. He cut 10 centimeters off one end and sewed it to the other end.

How long is the scarf now? (80 cm)

3. Mr. Mudanza had a hose that was 6 meters long. He made it a meter shorter at one end, and then he made it 2 meters longer at the other end.

How long is the hose now? (7 m)

4. Mr. Sleeby lives a block and a half from his toy store. He walks to the store in the morning and walks home in the afternoon.

How far does he walk altogether? (3 blocks)

5. Willy has 10 marbles. Seven of them are little.

How many of them are round? (all 10)

6. Ferdie and Manolita wanted to buy a bag of popcorn together, but it costs 15 cents. "I'll pay a dime and you pay the rest," said Ferdie.

Is that fair?
How much would Manolita have to pay? (5¢)

7. There are 4 doors in Mr. Mudanza's house, and each one had a doorknob. But Mr. Mudanza changed that. He put an extra doorknob on each door.

How many doorknobs are there now? (8)

8. Ferdie was taking 10 bottles to the store, but he dropped 9 of them.

How many good bottles are left? (Can't tell.)
What do you need to know? (how many broke)

9. Remember, Portia lives on the second floor.

How many floors down does she have to go to get to the basement? (2)

10. Mr. Tomkins drives a city bus. One day 4 people got on at his first stop. At the next stop, 1 person got off. At the next stop, 2 people got on.

How many passengers were on the bus then? (5)

11. Manolita belongs to a Brownie troop. There were 10 Brownies the first year. The next year 5 Brownies moved away, but there was 1 new Brownie.

How many Brownies were in the troop then? (6)

12. Mr. Nosho used to weigh 10 kilograms more than Mr. Sleeby, but now Mr. Sleeby weighs 15 kilograms less than Mr. Nosho. [This one is tricky. Read it again.]

What could have happened? What else? (Mr. Nosho gained weight, or Mr. Sleeby lost weight.)

Mr. Mudanza Changes Houses

Trouble in the Garden

19

Trouble in the Garden

In the springtime Ferdie went out to the country to help his grandfather plant his garden. The first thing that Grandfather wanted Ferdie to do was dig holes to plant some little trees in. "Please dig the holes exactly 15 centimeters deep," said Grandfather.

Ferdie grabbed a shovel and dug the holes in a hurry. "I'm done, Grandfather!" said Ferdie. "What's next?"

"We'll see," said Grandfather. He took a ruler and measured the holes. They were all 20 centimeters deep.

Is that what they should have been? (no)
Were they too deep or not deep enough? (too deep)

"Nuts," said Ferdie, "I made the holes too deep. Now I'll have to start over and dig new ones."

"I think there might be an easier way," said Grandfather.

Can you think of an easier way? (put some dirt into the holes)
How much dirt should Ferdie put into the holes to make them right? (enough to fill 5 cm)

Grandfather showed Ferdie how to pack 5 centimeters of dirt into the holes so they would be just 15 centimeters deep. Then he gave Ferdie some onions to plant. "Please plant them 10 centimeters apart," said Grandfather.

Ferdie took the onions and quickly planted a row of them. "I'm done," he said. "What's next?"

"Take your time," said Grandfather. He took his ruler and measured how far apart the onions were. "These onions are 20 centimeters apart," said Grandfather. "I asked you to plant them 10 centimeters apart."

How far apart is 20 centimeters? Show with your hands. [Demonstrate.]
How far apart is 10 centimeters? Show with your hands.

Trouble in the Garden

"Nuts," said Ferdie. "I'll have to pull out the onions and start all over."

Does he really have to pull them out and start over? (no)
What else could he do?

Grandfather showed Ferdie how he could stick another onion between every 2 onions that he had planted. Then the onions would all be 10 centimeters apart.

For his next job, Grandfather gave Ferdie a package of radish seed. "Please spread this out so that the whole package makes 1 row," said Grandfather.

Ferdie tried to be careful this time. He spread the radish seeds out slowly with his fingers. But when he finished the row he still had half the package left. "I did it wrong," Ferdie moaned. "I was supposed to use up the whole package and I used up only half of it. Now what do I do?"

Can you think of any way he can make it turn out right?

"No problem," said Grandfather. "Just go over the row again and plant the other half of the seeds."

"Now for your last job today," said Grandfather, "I'd like you to plant 4 rows of beans—exactly 4. I have to go back to the house, so I hope you can do this job all right by yourself."

"Don't worry," said Ferdie.

Do you think Ferdie will do it right? Keep track of what Ferdie does, so you can tell.

Ferdie planted 1 row of beans, then another row and another row. "Oh, my!" said Ferdie. "I can't remember how many rows I've planted."

Do you know? How many? (3)

Ferdie didn't remember that he had planted 3 rows. "I just can't remember," said Ferdie. "I guess I'll have to start over."

Is there any way Ferdie could find out how many he'd planted? (by going back and counting)

But Ferdie didn't go back and count how many rows he'd planted. Instead he went ahead and planted 4 *more* rows of beans. When Grandfather came back from the house he said, "That looks like a lot of bean rows."

How many rows did Ferdie plant altogether? (7)

"I lost count," said Ferdie. "I planted some rows, and then I had to start over and plant 4 more."

"Well," said Grandfather, "it looks as if we're going to have 7 rows of beans instead of 4. I hope you like beans, Ferdie, because you're going to be eating a lot of them!"

Trouble in the Garden

Problems

1. Ferdie is trying to help by pulling weeds in his grandfather's garden. Every time he pulls up 3 weeds he pulls up a bean plant by mistake. Ferdie has pulled up 12 weeds.

Can you figure out how many bean plants he has pulled up? (4)

2. "That's a tall pear tree, Grandfather," said Ferdie.

"I know it is," said Grandfather. "I climbed up 6 meters in it and then I could just reach the top with a 3-meter pole."

Can you figure out how tall the pear tree is? (Not exactly, but it's more than 9 m.)

3. Mr. Mudanza had a picture that was 20 centimeters high and 30 centimeters wide. He cut a centimeter off one side of the picture and a centimeter off the other side. Then he cut a centimeter off the top and a centimeter off the bottom.

What size is the picture now? How high? How wide? (18 cm high, 28 cm wide)
Do you want to hear the problem again?

4. "When will I ever get done painting these Easter eggs?" asked Ferdie. "It took me an hour to paint the first 2 and an hour to paint the next 2." He has 4 eggs left to paint.

How many more hours will it take him? (2 hours, at the rate he was going)

5. Portia walked 3 blocks away from home. Then she noticed that she had dropped her scarf, so she walked back 1 block to get it.

How far from home was Portia when she picked up her scarf? (2 blocks)

6. The Noshos live 2 blocks from the post office. The Breezys live 1 block farther away from the post office than the Noshos do.

Can you tell how far the Noshos live from the Breezys? (no)
What are some possible answers? (1 block, 5 blocks)

7. Mr. Mudanza had a square handkerchief. He cut the handkerchief in half, going from one corner to another corner.

How many pieces of handkerchief does he have now? (2)
What shape are they? (triangles)

8. The Mudanzas' bathtub used to be 45 centimeters deep, but Mr. Mudanza filled in the bottom of it with cement 10 centimeters deep.

How deep is the tub now? (35 cm)

9. The bananas cost 18 cents, and Mr. Nosho got 2 cents back in change.

How much money had Mr. Nosho given the grocer? (20¢)

Trouble in the Garden

10. Portia has a stack of books that is 8 centimeters high. Her friend Mark has a stack of books that is 11 centimeters high.

Whose stack of books is shorter? (Portia's)
How much shorter? (3 cm)

11. Willy was growing a flower and a weed in his flowerpot. Last week they were both 20 centimeters high. This week the weed is 24 centimeters high and the flower is 18 centimeters high.

What could have happened? (The weed grew; something made the flower shorter—the stem broke; the bloom was picked; and so on.)

12. Ferdie and Portia and Mark and Willy were drawing pictures. Portia was the second one to finish, Ferdie was the third one to finish, and Mark was the fourth one to finish.

What can you figure out about Willy? (He was first.)

Trouble in the Garden

20

How Deep Is the Water?

Note: The picture that goes with this story is best shown halfway through the reading.

Mr. Sleeby did not go fishing very often, because there were too many things to remember and he was always forgetting something. One day Ferdie went fishing with Mr. Sleeby and they tried to remember everything. They remembered to take fishing poles and string and worms.

Did they forget anything? (yes, hooks)

They got into a boat and rowed out on the lake to a place where Mr. Sleeby said there were sure to be some fish. But when they were ready to start fishing they discovered that they had no fishhooks.

"I did it again," said Mr. Sleeby. "I'm sorry. I'm afraid this fishing trip won't be much fun for you."

"That's all right," said Ferdie. "I like just being out here on the lake. I wonder if the water is too deep to stand in."

How could Ferdie find out?

Ferdie's fishing pole was exactly as long as Mr. Sleeby's height. Ferdie pushed the pole straight down in the water until it touched bottom. There were still 30 centimeters of the pole sticking out of the water.

Is the water too deep for Mr. Sleeby to stand in? (no)

"The water isn't too deep for the fishing pole. So it isn't too deep for me!" said Mr. Sleeby. And he jumped into the water and stood on the bottom.

How much of Mr. Sleeby was out of the water? (30 cm)
What part of him?

Only Mr. Sleeby's head was out of the water, so his clothes were all wet. "You'd better climb back into the boat," said Ferdie.

87

Note: This is a good point at which to show the picture that goes with the story.

"It might tip over," said Mr. Sleeby. "I think I'd better wade back to shore."

"But how will I get back?" asked Ferdie. "I don't know how to row."

"I'll push the boat," Mr. Sleeby said, and he gave the boat a push. It moved ahead 4 meters through the water.

"This is fun," said Ferdie. "I wonder how far it is back to shore."

Mr. Sleeby kept shoving the boat. Every time he shoved the boat, it went ahead 4 meters. Ferdie counted and discovered that Mr. Sleeby had to shove the boat 10 times before it got to shore.

Can you figure out how far they were from shore? (40 m)
How far did Mr. Sleeby have to push the boat? (40 m)

When they were back on land, Mr. Sleeby was dripping and shivering. "By the way," said Ferdie, "why did you jump into the water with all your clothes on, Mr. Sleeby?"

"I remembered how deep the water is," said Mr. Sleeby, "but I forgot how wet it is."

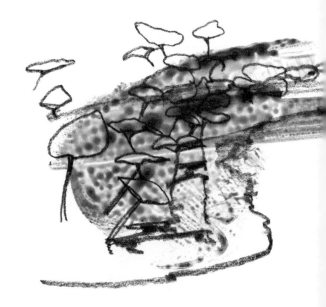

How Deep Is the Water?

How Deep Is the Water?

Problems

1. The next time Ferdie and Mr. Sleeby went fishing they caught 5 fish altogether. Ferdie caught 5 of them.

How many fish did Mr. Sleeby catch? (zero)

But Mr. Sleeby did catch 4 weeds, 2 tin cans, and an old bicycle tire.

So who caught the most living things? (Ferdie)

2. Do you know how Mr. Sleeby gets the toys he sells in his toy store? He has to buy them. So naturally he has to sell them for more than he paid. That's how he earns his money. If he buys a toy for $2, he sells it for $3. If he buys a toy for $5, he sells it for $6. If he buys a toy for $1, he sells it for $2.

If Mr. Sleeby bought a toy for $4, how much would he sell it for? ($5)

3. "How much is that yo-yo?" asked Ferdie.
"Forty cents," said Mr. Sleeby. "But since you're such a good friend, I'll sell it to you for half that price."

How much will Ferdie have to pay for the yo-yo? (20¢)

"If they're half-price, I'll take 2 of them," said greedy Ferdie.

How much will he pay for 2 yo-yos? (40¢)

4. The cuckoo clock in the Noshos' house goes "Cuckoo" once every hour. One day when the Breezys dropped in to visit, they heard the clock go "Cuckoo" just as they came in the door. The next time the clock went "Cuckoo" was when they were just leaving.

Be careful! Can you figure out how long the Breezys were there? (1 hour)

5. Willy has 12 cents, Portia has 9 cents, and Ferdie has 14 cents.

One of them has a dime. Who could it be? (Willy or Ferdie)
Could it be Portia? (no)
Why not? (She has less than 10¢.)

6. Mr. Mudanza used to have 26 teeth in his mouth, but he changed that. He went to a dentist and had 2 false teeth put in.

How many teeth does he have in his mouth now? (28)

7. Loretta the Letter Carrier bought 10 cents' worth of peanuts. She gave 5 peanuts to Ferdie and 5 peanuts to Portia.

How many peanuts does she have left? (Can't tell.)
What do you need to know? (how many peanuts she bought)

8. Mark ran around the block in 4 minutes. Manolita ran around in 6 minutes, and Ferdie did it in 5 minutes.

Who is fastest? (Mark)

How Deep Is the Water?

9. It costs a quarter to ride the bus. "Guess I'll have to walk," said Mr. Sleeby. "I have only 30 cents."

Is he right? Why not? (30¢ is worth more than a quarter.)

Finally Mr. Sleeby figured out that he did have enough money to ride the bus. He had 2 dimes and 2 nickels. Remember, it costs 25 cents to ride the bus.

What will Mr. Sleeby have left after he pays for the bus ride? (a nickel)

10. Mr. Sleeby was going to visit Willy, who lives 3 blocks away. Mr. Sleeby forgot and walked 1 block past Willy's house. Then he remembered where he was going and walked back to Willy's house.

How far did he walk altogether? (5 blocks)

How Deep Is the Water?

Correlation of the Thinking Story® Book and Teacher's Guide for Real Math™, Level 1

Lesson Number	Storybook Component of Lesson
1	Story 1, "How Many Piglets?"
2	2 of the problems following story 1
3	2 of the problems following story 1
6	2 of the problems following story 1
8	3 of the problems following story 1
9	Story 2, "Willy in the Water"
10	3 of the problems following story 2
12	2 of the problems following story 2
13	Reread story 2, "Willy in the Water"
14	1 of the problems following story 2
15	2 of the problems following story 2
16	Story 3, "Mr. Sleeby's Party"
17	2 of the problems following story 3
18	3 of the problems following story 3
19	2 of the problems following story 3
20	2 of the problems following story 3
21	Story 4, "It's Not So Easy," Said Mr. Breezy"
22	2 of the problems following story 4
23	2 of the problems following story 4
24	2 of the problems following story 4
25	2 of the problems following story 4
27	Story 5, "Mr. Sleeby Goes Shopping"
28	2 of the problems following story 5
29	2 of the problems following story 5
30	2 of the problems following story 5
31	2 of the problems following story 5
37	1 of the problems following story 5
41	Story 6, "Exactly What to Do"
42	3 of the problems following story 6
45	3 of the problems following story 6
46	3 of the problems following story 6
47	Story 7, "Willy Looks in the Mirror"
48	3 of the problems following story 7
49	3 of the problems following story 7
51	3 of the problems following story 7
52	Story 8, "Mr. Mudanza Builds a Better Tree"
53	4 of the problems following story 8
54	3 of the problems following story 8
55	3 of the problems following story 8
57	Story 9, "Mark Builds a Birdhouse"
59	4 of the problems following story 9
60	5 of the problems following story 9
62	Story 10, "Mr. Sleeby Buys a Candy Bar"
63	5 of the problems following story 10
71	4 of the problems following story 10
72	Story 11, "Mrs. Nosho's Fish Stories"
73	4 of the problems following story 11
74	2 of the problems following story 11
76	3 of the problems following story 11
77	4 of the problems following story 11
78	Story 12, "Manolita's Magic Number Machine"
80	3 of the problems following story 12
83	4 of the problems following story 12
84	4 of the problems following story 12
85	Story 13, "Mr. Mudanza Makes Lunch"
86	4 of the problems following story 13
87	3 of the problems following story 13
89	2 of the problems following story 13
90	Story 14, "Manolita's Magic Minus Machine"
91	3 of the problems following story 14
92	3 of the problems following story 14
94	4 of the problems following story 14
95	Story 15, "How Mrs. Nosho Doubled Her Money"
100	4 of the problems following story 15
102	4 of the problems following story 15
103	2 of the problems following story 15
104	2 of the problems following story 15
105	Story 16, "The Third House on Fungo Street"
106	4 of the problems following story 16
107	4 of the problems following story 16
108	4 of the problems following story 16
110	Story 17, "The Lemonade War"
111	5 of the problems following story 17
112	4 of the problems following story 17
113	2 of the problems following story 17
114	2 of the problems following story 17
115	Story 18, "Mr. Mudanza Changes Houses"
117	4 of the problems following story 18
118	4 of the problems following story 18
119	4 of the problems following story 18
121	Story 19, "Trouble in the Garden"
122	3 of the problems following story 19
123	2 of the problems following story 19
124	2 of the problems following story 19
125	2 of the problems following story 19
126	3 of the problems following story 19
127	Story 20, "How Deep Is the Water?"
128	3 of the problems following story 20
129	3 of the problems following story 20
130	2 of the problems following story 20
131	2 of the problems following story 20

Note: Some of the lessons suggest repeating a problem done in an earlier lesson if there is time.